Turning Your GREAT IDEA into a GREAT SUCCESS

Entrepreneur Judy Ryder tells how to develop, license, protect, & promote your product idea

JUDY RYDER

PETERSON'S/PACESETTER BOOKS
PRINCETON, NEW JERSEY

Dedicated to my loving family:
Michael, Mark, and Jessica

Turning Your Great Idea into a Great Success is published by Peterson's/Pacesetter Books.

Pacesetter Books is a trademark of Peterson's Guides, Inc.

Library of Congress Cataloging-in-Publication Data

Ryder, Judy.
Turning your great idea into a great success : entrepreneur Judy Ryder tells how to develop, license, protect & promote your product idea / Judy Ryder.
p. cm.
Includes index.
ISBN 1-56079-462-3
1. Inventions. 2. Inventions—Marketing. I. Title.
T339.R93 1995
608′.068′8—dc20 94-45493
CIP
Rev.

Book design by Kathy Kikkert

Printed in the United States of America

10 9 8 7 6 5 4 3 2 1

ACKNOWLEDGMENTS

Though I have countless people to thank in my education as an inventor, I must begin with the people most responsible for supporting my efforts to put thoughts on paper and write this book.

I never would have considered a book, nor found my way through the mystery of publishing, without the encouragement and guidance of Terry Sherf. Terry is not only a publishing guru, but a one-person cheerleading squad. Once I began, the only other human being to read these words was Marcia Steil, whose own expertise with the written word not only smoothed my rough edges, but has many times left me with tears in my eyes and a stitch in my side. To keep me going, I have a friend that anyone would be lucky to find in a lifetime. Nancy Hicks not only encouraged me, but kept better tabs on my deadlines than I did.

Many thanks to Carol Hupping and everyone else at Peterson's for allowing me considerable extra time to get this manuscript completed, after the Northridge earthquake so rudely interrupted my concentration.

My deepest thanks go to my husband, Michael, for always understanding my tendency to get carried away with ideas, and even though I didn't always like it, for always giving me an honest opinion when asked.

Aside from the book, thanks must also go to my mother-in-law, Monica Ryder, for unquestionable support at every turn and for sewing Little Shirt Anchors until she could cut, turn, pin, sew, and snip in her sleep. To Robert Ryder for assisting his wife whenever asked. To my brother, Randy Pausch, for being the best stripper in Southern California. (That's another word for lithographer.) To my mother, Merle Pausch, for support I can never repay. To Sue & Mel Munman for trusting me, for taking a chance, and for giving new meaning to "long-term loan." To Fonda Stevenson and her husband, Robert, for always being there.

Last but not least, thanks to my children, Mark and Jessica. They are an inspiration for everything I do in my life.

CONTENTS

INTRODUCTION

So you've come up with a great product idea—now what? How do you turn your great idea into a great success? How do you get it off the ground, get it made and sold? Whether your dreams are of wealth and celebrity status, newfound respect from your family and friends, or simply the chance to be your own boss, they won't come true until you take that first step. But what *is* that first step?

That was the question puzzling me in 1986, when I decided that a simple gadget I had invented to help dress my baby could be developed into a successful product: the Little Shirt Anchor.

It came about innocently enough. I had just had a baby girl, and a friend gave me a box of beautiful dresses for her. To keep the baby's chest and tummy covered during the cold weather, I put a T-shirt on under her dress. That did no good at all; the shirt was constantly riding up under her arms.

After struggling with this problem for awhile, I had an idea. Although I wasn't much of a seamstress, I was a determined mother. So I stitched together a contraption that would hold down my baby's undershirt—and the Little Shirt Anchor was born. Never again would any child have to live with a bare tummy.

Perhaps you could tell a similar story about your own invention. You struggled with a task once too often, then you took the time to invent a solution. Such solutions are what I call products of use. They are not fad items; no Cabbage Patch Dolls or Pet Rocks in this bunch. These are products that serve a purpose, that will endure the test of time.

It is important for rookie inventors to understand the difference between these two types of products right off the bat. We all dream of instant success with our inventions, but products of use are seldom "instant successes." Get ready to be patient. You don't get something for nothing, and that applies to your invention.

Does that sound like a preparation for disappointment? I don't mean it to be. I have written this book not to discourage you but to make you aware of what you may realistically expect as you proceed to turn your idea into a successful product. I don't believe in schemes that promise you can buy a house with no money down, and I don't believe in products that claim you can lose weight while you sleep. I do believe that with hard work, determination, and a little insider knowledge, you can successfully get your product idea to market. You will be unstoppable.

We inventors all feel certain that we'll strike it rich with our great idea. We've searched the marketplace and haven't found anything like it. Family and friends who have seen our product have reinforced our certainty. Our belief in our product, along with this reinforcement, motivates us when the going gets really tough.

If you decide right now that this is it, that you're going to achieve great success with your invention, then you have discovered the first ingredient in marketing and selling your idea: confidence.

The second important ingredient is determination. Have you ever heard the saying, "Good ideas are a dime a dozen"? Well, it's true. I now work as a consultant to other inventors, and I've heard my share of fantastic ideas—but that's not all it takes.

The big question is, how prepared are you to develop your idea? How inspired are you to market it? Are you courageous enough to put your money (or someone else's) where your mouth is? To really get out there and give it your all?

As you strive to make your invention a success, you'll be working toward a goal that is attained by only a select few. Do you have what it takes? If only there were some way to predict the future, to see the outcome before investing all the time and money, the blood, sweat, and (dare I say) tears.

Am I making this sound like it's going to be harder than you thought? Good! Because it is.

My delusions of becoming filthy rich off my first product are gone, though I have enjoyed a profit. But more than that, I've had the reward of knowing I accomplished something that *millions* of people have tried to do and very few people have succeeded at. I achieved things as a novice that surprise even veterans in the field. Long before I was a financial success, I labeled myself a success.

While it is a struggle, the sense of accomplishment is comparable to nothing else. As you go further along in this arena, you will meet other people who are struggling, and you will watch many of them fail. It will dawn on you that you really are doing an extraordinary thing. Even if you don't end up with great wealth, it will be an experience you will never forget. Your achievements will be a feather in your cap for the rest of your life.

When you're just starting out, so many things can happen; so many decisions have to be made. It sounds like a game show ("I'll take curtain number one"), doesn't it? Well it is, sort of.

To be ready to make those decisions, you must have confidence in your abilities. Knowledge helps build self-confidence, and reading this book is a great place for you to start. You will learn everything from how to price your product to how to spend your advertising dollars. It will give you a little peek behind the curtain before you have to choose.

I suppose if you've read this far, you're hooked. I realize you may skip around from chapter to chapter, and I hope you do. Please use this book as a reference manual whenever you are faced with a situation you are not quite sure how to handle. I've done my best to cover each subject thoroughly; where I haven't provided enough detail, I've provided the names of places and people to refer to for more information.

Since inventing my first product I've had many opportunities to grow and educate myself by working with some of the most successful business people and companies in this country. I am now, and have been for some time, ensconced in the corporate world; my days are filled with board meetings, sales strategies, and product line reviews.

There have been many product launches of mine between the time of my first—the Little Shirt Anchor—and my days now in the corporate world. But I've deliberately chosen to use examples from the Little Shirt Anchor throughout this book so that you have the continuity of one product's development, rather than pieces of several.

This book is not for you if you've been in business 15 years. This book *is* for you if you have little or no experience, little or no background—just a burning desire to get your product out there for people to buy and use.

The key is to amass all the knowledge you can, then make your own decisions, based on facts and instinct. Others, including me, can only give you so much advice. You are the one who has to live with the outcome.

Sometimes as I was writing this book, I imagined my readers feeling just like I did when I started out: anxious, happy, certain that greatness was just around the corner. I wish I could speak personally to each and every one of you—if not to give advice, then simply to provide inspiration.

You'll go through periods when no one can stop you; then you'll go through periods when all you can think is "What have I done?"

Just remember that it's very normal to feel unsure. But the more good decisions you make, the more confidence you will have in yourself. And remember: You are not alone; you are part of a special breed.

Now go out there and make your dream come true!

CHAPTER ONE

GETTING STARTED

When I invented the Little Shirt Anchor, I started thinking very big very fast—dreams of fame and fortune were all my head had room for—never mind reality! But I learned my lessons; and as your guiding spirit for the duration of this book, I'm going to make sure you're exposed to no end of reality. Of course, there's still plenty of room for dreams; without them you may never have invented anything in the first place. But to fully realize your dreams, you need to start your business off on solid ground.

Home Base

First things first. As a new inventor about to launch a business, you're about to make some big changes in your life. If you're married, and your spouse isn't going to be involved in the business, now's the time—before things get crazy—to sit down with him or her and have very frank and open conversation. What expectations do you and your spouse have? How much time do you expect to spend on this business? How much time does your spouse expect you to spend? How much time and money are you willing to invest before you see any profits? If you're a parent, like I am, you can almost certainly anticipate moments of frustration when, for the fourth day in a row, you call out for pizza and the kids can't find any clean socks for school.

> **If you're a parent, you can anticipate moments of frustration when, for the fourth day in a row, you call out for pizza and the kids can't find any clean socks for school.**

Whatever your share of duties at home, rest assured that you will fall behind. You will, from time to time, need a hand from your spouse. If this is going to be a problem, you should know it ahead of time. And have you thought about how you will handle a spouse who becomes a bit jealous, for example, of the interviews you are asked to do—at trendy restaurants, perhaps—when you're trying to publicize your product? And if lunch is going to cause bad feelings, what about overnight trips or an interview or two on TV or in the local paper? To make matters worse, you won't even be bringing any money in, at least at first.

These days a lot of my time is spent counseling other inventors, and half of my clients, at one time or another, have told me about troubles related to their business and their spouse. This is a very real obstacle. The best thing I can tell you, again, is talk problems out, whenever possible, *before* they reach crisis stage.

Separating Work and Home

In addition to time spent away from home, there's the problem of business intrusions within your home. You need to think about what area of your house you can dedicate to your work—an area that will cause minimal disruption both for you and for the rest of your family. If you're like most small businesspeople I know, you'll be bursting at the seams before you're financially ready to move your business out of the house and into a true workspace.

When I started building my Little Shirt Anchor business, my office, my shipping and delivery area, and my storage space were all in one location: the living room, which was about 12 by 25 feet. Our family room, which measured 15 by 30 feet, was the "sweatshop." It contained the three industrial bar tack machines I had rented, along with two straight sewing machines. From nine to five, Monday thru Friday, this sweatshop was full of my workers: three people full time, and temporaries when the orders piled up. We had Little Shirt Anchors stacked against every wall from floor to ceiling, in various stages of completion; bags of elastic and garter grips; yards and yards of fabric; and reels of Velcro, along with all the workers snipping and stuffing. Sometimes my husband would get very upset because there was no place in the house to sit and have his morning coffee!

Thats may be your reality for the time being, too. Just remember it won't always be that way.

RULES OF THE GAME

Now for the basics of the business world. But wait—maybe you're certain your product idea is so outstanding that the rules just don't pertain to you at all. People who see your product *love* it. It is so different and innovative that Walmart or Toys 'R Us will just roll over and play dead for the chance to carry it.

Guess again.

Set Your Own Goals

It's easy to get in a tizzy thinking about $500,000 checks and what you're going to do with all that money. I certainly had dreams of growing rich after I first invented my Little Shirt Anchor. You see, I had heard a story about a young man who lives in my neighborhood. He came up with an idea for a protective covering for his watch, kind of like a little cross made out of plastic. He made a prototype (sample), and took it to a watch trade show. The Swatch company saw it, fell in love, flew him to their offices, and proceeded to make him a very rich young man. Rumor had it that he received $250,000 royalty checks every three months for his invention, the Swatchguard. I figured my product would sell for more than a Swatchguard, so I could potentially make more than that!

Don't set goals based on what someone else has accomplished. Set your own goals—and be realistic.

Looking back now, I'm a little surprised at my own ignorance. (Then again, I also buy lottery tickets.) The Swatchguard story was the only measuring stick I had for success as an inventor, so I used it on my product. (And let me reiterate this was the "story" going around not based on fact.) That is kind of like planting cucumber seeds and expecting to grow an apple tree.

The Little Shirt Anchor and the Swatchguard aren't the same sorts of inventions. The Swatchguard is a fad item, designed to be purchased by the portion of our population who spends more than any other group. It is not a product of use—an item, like the Little Shirt Anchor, that fills a need and has staying power. Oh, I know the Swatchguard is supposed to protect your watch's face, but most people buy one because it looks "cool." A person does not look at that product and say, "I really need one of those."

Another problem with my comparison is that I really didn't know how much of the rumor about the Swatchguard inventor's royalty checks was true. But true or not, it's simply not wise to set goals based on what you read or hear someone else has accomplished.

When you're starting out, set your own goals—and be realistic. If you measure your success only by the size of your bank account, chances are you will be spending many months feeling like a failure, which you're not. The adventure you are beginning will take time. You'll need to be patient and understand that not all things are going to come out the way you planned or as soon as you planned. Give yourself a break.

Fads vs. Products of Use

It may seem that a product of use would have the advantage over a fad item in the marketplace, but unfortunately it works the opposite way. Statistics show we are more likely to purchase something based on emotion than based on need. We see it all around us in advertisements. Pet food, soft drinks, you name it—manufacturers appeal to our emotions to encourage us to buy their products.

A product of use is not an impulse buy. Usually, when we encounter a situation in which we really need a certain item, we file it away in our memory banks, and then, after encountering that situation a few more times, finally break down, go to the store or catalog or whatever, and buy the item.

So deciding to buy a product of use is very different from deciding to buy a fad product. And that was something else I had neglected to consider when comparing the Little Shirt Anchor to the Swatchguard: Fad items may sell in one rapid spurt; sales of my product were going to build *slowly*.

Okay, your expectations are now sufficiently down to earth. Ready for the next step?

A Lesson Learned: Don't Let Little Setbacks Get You Down

I have told many clients as they begin this journey that for every 100 possibilities they work on, expect one to come through. From manufacturing, raising capital, press releases, product submissions, and every other aspect of inventing a product, you must realize that it is very common to experience more disappointments than days of celebration.

When I first started with my Little Shirt Anchors, someone had given me a bottle of very expensive champagne. She told me to keep it in my refrigerator until my big day came. Well, I kept it in my refrigerator for almost four years. In that time I had JCPenney as an account, I licensed the product two times, and I appeared on countless TV talk shows, and in magazines and newspapers across the country.

But none of those things was The Big One. None of those things made me feel like opening the champagne. You see, it is a series of successful events that will make you successful. And it's a series of disappointments that will give you the knowledge and drive you need to succeed.

With the JCPenney account came an extra $4,000 in packaging expenses for me, because Little Shirt Anchors had to be put in blister packs *(see page 111)**. My first license agreement was with a company that went out of business one year later. And all the wonderful TV, radio, and print interviews meant travel expenses and time away from home.*

Learn to pat yourself on the back, and don't let the rejections and the little setbacks get you down. They are part of the process. What counts is that, added together, you and your product are making progress and you're learning from your mistakes as you go.

Laying the Groundwork

When you come up with a great idea for a new product, its greatness is largely measured by its uniqueness. Don't think that just because you have never seen a product before, it doesn't exist. You might just be looking in the wrong place. Companies distribute products in a variety of ways. Some just sell to mail-order catalogs. Others have exclusive arrangements with one particular store chain. Still others may sell their products directly to other manufacturers rather than to the public.

Your product's uniqueness will be a key factor when you attempt to license it, try to raise money to get started, or simply try to evaluate whether you should continue with the project.

It is equally important to determine whether any product similar to yours has ever been on the market. If something was around five or ten years ago that is no longer available, you must find out why. Was the price too high? Did it work incorrectly or fall apart? Was the supplier having problems? Did the public just not like it? Were there safety problems or was the product recalled?

DOING YOUR HOMEWORK: MARKET RESEARCH

As you start thinking about the prospects for marketing your new product, one of the first things you should do is become familiar with the industry you'll be working in, be it fabric and sewing supplies, home furnishings, or men's toiletries. There are several ways to study up on the whos, whats, and hows of your chosen field.

Catalogs

In this day and age, a very important part of distribution for many products is mail order, and consumer mail-order catalogs that carry products of the same nature as yours are a valuable source of information. Start by asking for a directory that lists mail-order catalogs at your public library.

Contact the customer service department for each catalog and ask to get on the mailing list. You may also try asking customer service reps if their company has ever sold any product similar to yours. (Don't worry too much about a busy person on the phone stealing your idea. You don't need to reveal any details. If such worries start getting the best of you, go directly to chapter four, to the section on

(text continued on page 15)

A Lesson Learned: Stay Focused; One Step at a Time

I recently was referred to a client through my banker. We met for lunch, and she began her presentation to me in a very organized fashion. She had artwork completed for her product. She had all her potential markets and potential offshoot markets. I don't want to divulge what her product idea was; I'll just tell you that it was essentially a gift item for kids.

She covered the grandparent market, had an idea to do an annual birthday gift program, was ready to approach major airlines such as Delta, hotel chains, mail-order catalogs, and stores. She already had her 800 number, she had financing in place and had ideas on how to customize the gift for several different age groups.

Halfway through the lunch, I was wondering to myself what this person could possibly need from me. She had all her manufacturing costs, had done her market testing, and had her patent and trademark done.

She also had a business background, so she had an understanding of wholesale and retail pricing, production, and marketing. As we continued with our meeting, it finally began to dawn on me: She hadn't done anything.

Well, let me rephrase that. She had done all the background work she needed. But she hadn't started. She hadn't sold even one of her product, even though she had been working on it for almost four years. She hadn't advertised. She hadn't asked a store if they wanted to buy some. She talked about it so much, for so long, imagining all the potential for her product, that she forgot to actually do it.

I know this all may sound strange and improbable to you, but believe me, that is just what she did. And it is something that almost anyone completely caught up in their anticipated success could do without even realizing it.

(continued on next page)

Sometimes we do so much planning in our own minds, we hype ourselves up so much by talking about our idea with our friends and family, sometimes attaching their great suggestions to our original plan, that to actually begin is something that seems overwhelming to us. Suddenly we begin to feel that unless we do all the things we have planned, we will be failing ourselves and our product.

How could the woman in this story actually go back and start with a tiny little ad in the paper, or try to get into her local stores, when she had already ''lived'' the reality of selling to Delta Airlines?

Somehow it feels like we are doing something wrong if we don't do it all at once. If we don't prepare our brochures to send to the catalog companies, and coordinate advertising to hit every potential market, and get our product in bulk packaging, and contact exporters, and add other products, and, on and on....

Stop! Keep focused.

Your product idea may be perfect for kids, and in a different package it might be just great for athletes. It may also be a wonderful premium, or giveaway. And, it may have offshoot markets, such as in the story here, of hotels or airlines. But don't try for each of these markets all at once! Start with one and when you have that going, go on to the next.

---

the affliction known as Inventor's Paranoia.) They may know of such a product that was available in a previous edition of the catalog.

As you acquire these catalogs, save them. They're a great reference source for you, and the catalog company itself may be a future customer for *your* product.

Trade Magazines

You should subscribe to any trade magazines that cover your product area. (Ask for back issues too.) These magazines are written for the people behind the scenes, not the public. People who work in or who own baby specialty stores, for example, read the two major trade publications, *Small World* and *Juvenile Merchandising*.

Buyers who specialize in juvenile products such as my Little Shirt Anchor review these publications for new products on the market, trade show information, and buying trends, plus mergers, acquisitions, and promotions within companies. All of this can be vital information for inventors ready to launch a new product. Your local library has books that list these publications and their addresses. Reading the trades will familiarize you with the "movers and shakers" in your industry. You'll also learn the particular lingo used, so that you'll be able to communicate in a more knowledgeable way.

Consumer Shows

A new product's greatness is largely measured by its uniqueness. Don't think that just because you have never seen a product before, it doesn't exist.

Keep a watchful eye out for consumer shows or fairs that are likely to carry products in your category. Attend them and gather all the information you can. You'll also want to visit every store in your area and scrutinize the shelves—specialty stores in particular, as they usually carry many products that cannot be found in mass merchant and department stores. If you find a salesperson who has worked in the store for some length of time, you may, again, want to ask whether the store has carried a product like yours in the past.

And again, you don't need to divulge any details about your product; try to ask general questions about the category of product. Find out as much as you can from everyone you talk to. Remember, you are no longer visiting stores as a consumer. You are a businessperson researching your industry.

Comparison Shopping

Once you have your information, what do you do with it? If there is or was a product already on the market, find out everything you can about it. Is your product different? I mean really different? If so, is it an improvement that seems to be obvious, to make sense?

Try to turn a negative into a positive. Perhaps the product already on the market doesn't work quite right but still sells fairly well. If that's the case, fix the problem with your product. Or find out if you can make a modification to your product idea, maybe even coming up with a new product in the process. That's called thinking smart.

Reading the trades will familiarize you with the "movers and shakers" in your industry and teach you the lingo that can make you sound more knowledgeable.

As you do your comparison shopping, keep in mind that you're comparing products, not people. Think in positive terms: Don't denigrate the other product; just make factual comparisons and note your own product's strong points. Don't get set to be angry at someone else who had your idea first. Don't convince yourself that your product is better just because it's yours. It will not benefit you in any way to harbor negative thoughts. If you start thinking positive now, you will radiate that attitude when you interact with others.

I can't possibly advise you on every scenario you might encounter, but I will give you two guidelines to follow:

First, try your best to be a nonbiased judge of the situation. Separate yourself from your identity as an inventor and be a businessperson.

Second, realize that you and your product, no matter how good it is, are not special enough to overcome all obstacles. Some are simply insurmountable. Success takes much more than a good idea, and sometimes even determination isn't enough. If you discover that you'll be competing with a major manufacturer, for example, you may be in trouble. Even if you make a product that is superior, the stores are still likely to buy from an established supplier rather than from a new inventor. Or you may find that a similar product was on the market five years ago and there was an injury or death associated with it. Maybe the Consumer Product Safety Commission has not yet banned the sale of this type of product, but you would be crazy

You and your product alone are not enough to overcome all obstacles. Success takes much more than a good idea.

to attempt to market anything similar. That's what I mean by an obviously insurmountable obstacle.

Competition is healthy—even for inventors. Unless your competitor is a major in the industry, having a rival simply means the market will accept your product more quickly, because you are both trying to sell the same idea; people are hearing from more than one source that this is a great new product. Use that to your advantage. If your competitor has one area really flooded, concentrate your efforts in another area. Competition leads to creativity. Take my word for it.

CHAPTER TWO

PROTOTYPES AND MARKET TESTING

So you've done your preliminary research, and you're feeling pretty confident that there is, indeed, a place in the world for your invention. Now it's time to put your act together and take it out on the road. In order to do that, you first must turn the idea in your head into a tangible, testable product—a prototype.

DEVELOPING A PROTOTYPE

You absolutely must make a sample of your idea, to see if what you have envisioned will really work. You need to check size and motion. Or stability, or materials—whatever pertains to your particular product. I can't be more specific than that, since I don't know what your invention is, but I can guarantee one thing: Whatever you have pictured in your mind will not work exactly as you have imagined.

If your invention is a game, you can't possibly write the instructions if you haven't played it several times and observed others playing it. Every individual handles a product differently, and you can't imagine every situation that may arise. That's why it's important to check and double-check your design. If you design a portable workbench that folds in half for an at-home mechanic, you may determine the height of the average garage door and build a beautiful sample, only to discover you didn't include the extra height for the casters on the bottom.

It's critical to make a prototype. Even the simplest ideas have to be tested before you can be sure you have a viable product.

Details, Details

The simplest things are the ones that will escape you. For example, an inventor I know developed a tube for dispensing caulk with a cap that was designed to push out the caulk. Until he made a prototype, he didn't realize that the cap would be too wide to act as the pushing device. I know, it's hard to imagine *you'd* make a mistake like that, but it happens all the time.

Have you ever undertaken a project such as refinishing a piece of furniture or putting together a model car, and found the instructions made it sound much easier than it turned out to be? Even though you knew in your head what was involved, when you got to work you suddenly realized you didn't have the right sandpaper or you needed two people to hold something in place while you put in the screw.

It's the same thing with your product. Will the Velcro actually stick? Will the plaster harden quickly enough? Are there enough suction cups? Will the peg fit through the hole? You just can't anticipate every detail until you've got a sample to work with. Even the simplest ideas have to be tested before you can be sure you have a viable product.

Putting It Together

How do you make a prototype? Well, it all depends on your product. You can do as I did when I developed a disposable baby bottle that expels the air automatically, you can buy an existing product, cut it up and add bits and pieces of other items. Or, you can get out the old sewing machine and stitch one up; or you can go to a craft or hobby store, hardware store, or electronics shop and buy bits and pieces to create your product. We are not interested in looks here, just in functionality. Be creative and use your imagination. For example, if you need a plastic part in a particular shape, perhaps you can make one out of clay. However, this homey approach won't work for all prototypes. If you're undertaking something more complex, you may want some outside help.

One option is to offer a small part of your business to someone who might want to invest in it by making a prototype for you.

For complicated sewing jobs, you can find a sewing contractor through your local newspaper and sewing shops; this is also a good way to get a pattern made if the person you hire has experience making patterns for commercial use. (A pattern for home use is very different from a pattern a sewing contractor would use in a production run.)

A plastic sample or prototype can be much more difficult to come up with. Some of my consulting clients have had local manufacturers make a prototype for them for a reasonable fee. (For information on finding manufacturers, see chapter five.) When I wanted a prototype of the baby bottle I mentioned earlier, I first bought a few copies of the existing product. Then I went to a plastic company, which had all kinds of tubing, rods, knobs, etc. I bought a little of this and a little of that and brought it home.

I talked my idea over with my husband, who was enthusiastic about it. (Not always the case: He sometimes gets tired of hearing about my latest new idea. I can't imagine why.) After hearing my thoughts and applying a few of his own, he used his power saw and some epoxy that works on acrylic, and tah-dah . . . I had my prototype. It didn't look anything like a manufactured product would look, but it demonstrated size, function, and assembly.

Another option is to offer a small part of your business to someone who might want to invest in it by making a prototype for you. You may want to ask at a local college or university that teaches a course in manufacturing. Frequently they like tc have a real product to use as a class project, and sometimes you can even get a mold or tooling out of the deal.

TAKING IT ON THE ROAD: MARKET RESEARCH AND TESTING

When you've got your prototype all stitched or glued or otherwise assembled, and you've tested it for basic functionality, it's time to put it to a tougher test—this time in public. The process is known as market testing, and it's an excellent tool for assessing your product and building your knowledge of your target markets, i.e., your future customers.

What is market testing? The term itself implies that you test the marketability of your product idea before it is actually on the market. The only way to do this is to expose the product, in one form or another, to potential customers and record their responses to it. Why do it? To gain valuable information about your product. Remember, the more you know in the beginning, the more effective you will be in all phases of development. If you want to license your product (see chapter six), for example, this will be great information to include in your presentation.

You may have participated in a market test or focus group at some time in your life. Market testing is done on all kinds of products, from dish detergent to motion pictures. A couple of years ago, for example, my daughter and I were paid $50 to watch a Batman cartoon with five other kids and their moms. The market testers asked the kids how they liked it; then they asked the moms how we liked it. None of the moms thought the cartoon was appropriate for kids. Later I saw it broadcast on TV at night, for adults. The whole test had taken less than half an hour, but obviously the production company gathered some valuable information about their cartoon.

Products are tested early on, while they're still being refined, and sometimes again in their finished form, once a package and name have been decided on. In the latter case the product is placed in a store setting, where sales and comments are closely monitored.

Market testing is an excellent tool for assessing your product and building your knowledge of your target markets—your future customers.

Sometimes market testing will reveal competitors you didn't know were out there. Your test may tell you what color or design to use. It may reveal that you've hit on a great idea, but no one is willing to pay over $5 for it. Maybe people would feel unsure about giving it as a gift because of sizing. You don't know what you're going to learn ahead of time—just remember what you don't know *can* hurt you.

Market Research on a Budget

When I started out to market the Little Shirt Anchor, I couldn't afford any sort of elaborate market research. I just made some samples and handed them out, along with some questionnaires; and simple as it was, it worked. I did the same thing with my shoelace invention and my games. These items were inexpensive enough for me to make several samples in exchange for receiving valuable feedback. That's a good example of something you can do right now, without spending a lot of money or wading through ten years of business school.

I am going to make the grand assumption that it is not possible for you to spend $25,000 on a market test. So here are some ideas on how to do a little testing on a reasonable budget:

Carefully determine what target audience is best for your product. You have to show restraint in your evaluation. I know as an inventor you may tend to feel "Everyone would want one!" But that's just not true—not for any product. All items have a target market. Is it a particular age group? Is it hobby enthusiasts? Is it pet owners? You must create a customer profile and make it slim. It is better to underestimate by a little than include the whole world.

There are simple types of market research you can do right now, without spending a lot of money or wading through ten years of business school.

The people you select for your test must be potential purchasers of the product; otherwise it's a waste of time. I certainly wouldn't go to a construction site if I wanted to get opinions on a new type of mop. I might, however, go to a shopping mall or child-care center. This is targeting your customer. Of course, I'm not saying that men don't mop; they just may not do it as often. So I would head for the sure thing.

Don't overlook the gift potential of your product. If you have a children's product, you may want to include grandparents in your market testing. If the product is a utilitarian item, you'll want to limit yourself to grandparents directly involved with the care of their grandchild(ren). If the product is more of a fad item, then you can broaden your test market to include gift-givers.

Think carefully about what questions you will include and how you will phrase them. I always follow a few simple rules: First, limit yourself to ten questions, even if you've thought up more. The average person will be overwhelmed at the prospect of filling out a long questionnaire and therefore may not answer accurately.

Second, use a multiple-choice answer format whenever possible. I've found that when people are asked to fill in a blank, a blank is what you usually get back. A couple of exceptions to this rule are questions meant to help determine the name of a product (i.e., "What word best describes the texture of the fabric?") and an overall "comments" question at the end.

Third, when asking about price, include two very different questions: "What would you expect this product to cost?" and "What would you be willing to pay?" My rationale for the first question: How many times have you seen something in a store or catalog and placed an estimated value on it, only to be surprised to find it really costs much more or much less? When the price is lower than customers expect it to be, many times the "bargain" light goes on in their minds, and an item they might not otherwise have seriously considered purchasing becomes a deal that's too good to pass up. The second question is rather self-explanatory. These two indicators—customer expectation and willingness—help you make important decisions about pricing—which influences decisions about packaging and placement as well.

All items have a target market. You must create a customer profile and make it slim.

A detail worth keeping in mind: When phrasing the multiple-choice answer to a question involving a number—e.g., "How much are you willing to pay?"—don't list the amounts sequentially—"$2 $4 $6 $8 $10"—or you will more than likely end up with many $6's, simply because it's the middle amount. People are automatically drawn to the middle. List your numbers out of sequence. It'll make your respondents think about it a little more, which is what you want. It's also usually a good idea to include the response "other ________" for those who think differently and want to express themselves.

What else to include? You may want to ask people where they might be apt to look for your product. Or what color(s) they'd like it to be. Or if they've ever seen anything like it. Or whether they would buy it as a gift. "On a scale from 1 to 10, rate this product" is always a good one. Use your imagination—but be brief and to the point.

Assessing the Answers

Just a word about analyzing the final answers you come up with. Statistical analysis is a skill that is learned over many months or years in school and many hours of practice. I could never explain in ten books all the different formulas used. But you can also use common sense. If you intended to sell your product for $25, for example, and your market-research respondents have indicated they'd be willing to pay an average of $3.95, you need to go back to the drawing board. Try to apply logic and see what you come up with. And don't be angry with negative responses. They are there to teach you something. Learn from them.

Most people enjoy giving their opinion, especially if they get something in return.

The Market Test in Action

Now you know what to ask; so how do you go about asking? Let me explain in more detail what I did with my Little Shirt Anchor.

I made 150 samples. Yes, it cost me some money, but it was well worth it: I wasn't willing to risk the money it was going to cost to start a business if I didn't know my product would really work and people would really use it.

I took my samples, along with the questionnaires I had made, self-addressed, stamped envelopes, and some quarters, down to the local mall. I also took my daughter in her stroller. I stopped potential customers—anyone with a baby—and asked if they would like to try a new product I had invented. I told them they could keep the product as long as they returned the questionnaire to me within two weeks. I also gave them a quarter for their child's piggy bank.

I think those quarters were what helped me get so many responses. A personal touch like that can really make the difference. In my case, it helped me get a lot of valuable information about sizing and colors, when my product was used, what benefits were most important, where customers would look if they wanted to buy one, and gift potential.

Note: I need to mention here that I opened myself up to a huge liability risk when I began market testing my Little Shirt Anchor. I had no product liability insurance, so I was personally liable if anything went wrong with one of my products. Please be aware that you need to get professional advice before providing your product to individuals, even if it's only a sample for testing.

I can give you another example of some market testing: When I invented my first game, "Shuffle," my partner Lois and I went to a Mexican restaurant that had an outdoor patio, about every other week. We went for dinner and usually had a margarita or two. We always invited different people to go with us; sometimes adults, other times just kids. We played the game hundreds of times, inventing new games in the process. We also observed our "guests" playing to see what excited them and what didn't. We began to refer to our evenings as "business meetings" and the Mexican restaurant became the "office." Once we established all the rules and variables to playing, I made several games and gave them to friends and family members that I knew liked to play games.

Creativity Counts

Of course, your product may be much more complex than my Little Shirt Anchor, and passing out 150 samples may not be an option. But you can still put on your thinking cap and get out there. Say, for example, you have a pet product, and you can only afford one prototype. Go to your local pet store and tell the manager you're an inventor from the area and you're a frequent customer in the store. Say that if you are permitted to ask the store's customers a few questions, you'd be happy to buy $1 or $2 gift certificates to hand out. If you get an okay, you'll want to create a streamlined questionnaire that people can zip through.

I have found that most people enjoy giving their opinion, especially if they get something in return. In one of my market testing schemes, I asked the manager of a local diner to pass out "invitations" to the customers as they were seated, asking them to take two minutes to watch a product demonstration in the corner of the restaurant, then fill out a questionnaire at their table while they ate. When they completed the questionnaire and returned it to me, they would receive a coupon for $3 off their next meal. The restaurant owner benefited because I paid for the $3 coupons and he got those customers to come back for another meal.

In a focus group, you should emphasize that you really want the truth about your product, not a "friend's" truth.

FOCUS GROUPS

Another way to gather responses is in a focus group. When I do market testing on a new product, I invite groups of 25 to attend each focus group, with the hopes of getting at least 18. I pay at least $20 per person and ask them to plan on spending between 30 minutes and one hour at the session.

During the focus group, the product is demonstrated, then participants are asked to fill in questionnaires. I ask participants not to speak about the product during questionnaire time so that their opinions won't influence others. Afterward, there is a group discussion about the product, during which I take notes. Then come refreshments.

Making a focus group run smoothly takes practice. You may want to try a mock group with a smaller number of people, perhaps people you know—as long as they qualify as potential customers. If you use people you know, your focus group should be blind—meaning no names on the questionnaires. You should also emphasize, before you start, that you really want the truth, not a "friend's" truth. Make it clear to participants that they would be doing you an injustice by lying. Still, if you use people you know you'll probably get some who'll try to protect your feelings, so figure this into your analysis.

Don't use a sales pitch during the group. When you talk about the product, discuss only its features, not its benefits. Don't announce, "This terrific little pouch in the front holds your wallet and keys, freeing your hands." Simply say, "It has a pouch in the front." You won't be doing yourself any favors by showing bias.

If all of this sounds like a bit too much to manage, you can always hire a company to do a focus group test for you. But it's not cheap. Costs vary greatly, but such a service usually starts at several hundred dollars.

CONSUMER SHOWS AND FAIRS

Shows, fairs, or expos, as they are sometimes called, can focus on anything from RVs to pet products to guns to baby products. The big ones are often held at sports arenas or convention centers; the smaller ones can be found in high school gymnasiums or hotel ballrooms.

I have attended many of these, both as a consumer and as a vendor or participant. As a vendor I only had one successful show, and that was working with someone else's product line. All the rest were duds. That doesn't mean they're going to be a waste of time for you, but there are many factors to consider before you get involved.

Shows Aren't for Everyone

Participating in a consumer show involves renting booth space, so you want to be fairly certain ahead of time that the show will be worth your investment. I would not attend a consumer show with a single product, because in most cases it's a waste of time, and you will not make your booth space fee back. Of course, there are exceptions to this rule: Once when I was attending a cat show representing my pet products company, It's Paws-ible, I saw a fellow selling a sticky roller you use to get the cat hair off your clothing, furniture, etc., and he seemed to be doing very well. My friend and I even bought one. He had an economical solution to a common problem, and it took off.

The Ground Rules

If you do decide to do a show, keep a few pointers in mind:

LOCATION, LOCATION, LOCATION. Try to get a corner booth in the center of the grounds. Avoid the booths around the edges, except near the entrance door or next to the restroom. And find out what kind of activities will be going on that day. You don't want to be next to the Dixieland band all day, yelling at your customers. Will you be next to a guy selling a book on how to train an elephant, with live demonstrations? Probably not a good thing; people will be in front of your booth, trying to see the elephant, not your product. Near the elephant is okay; you can grab the overflow. Next to the elephant is not okay. And it's not just elephants you should be leery of. Any animal, even worms, can be attractive. I was at a bird show once where a booth featured open displays of live worms wriggling around in wood shavings. Well, every kid in the place mobbed this booth all day long. You couldn't even walk down the aisle.

FIND OUT IF THE SHOW WILL FEATURE DRAWINGS, GIVEAWAYS, OR RAFFLES. Can you donate? Mentioned in A Lesson Learned, p. 31, I raffled gift certificates for a hot air balloon flight at my Little Shirt Anchor booth; however, some shows don't allow individual drawings. Check the rules before you make plans.

ASK ABOUT COMPETITION. You need to know if some other brainy inventor has come up with a product similar to yours. If so, good luck. (See my tale of woe in A Lesson Learned: Know Your Competition.)

BE SURE TO CALCULATE YOUR COSTS ACCURATELY. Along with the booth rental cost, you must budget for parking, gasoline, hotel, food, and booth design costs. You really have to do well to come out ahead in a consumer fair, and I can tell you it doesn't happen very often.

(text continued on page 33)

A Lesson Learned: Know Your Competition

After I had achieved some success with my invention, the Little Shirt Anchor, I didn't care about competition—there are, after all, enough babies (customers) to go around. However, I didn't always feel that way. In fact, when I discovered I had a competitor right out of the chute, I was devastated. I couldn't believe someone else could have come out with my idea so quickly.

My competitor was a woman who happened to live about 15 minutes from my house. I truly believe, though, she had never seen or heard of my product before she made hers. We just both had the same idea at the same time.

When we were both starting out, we attended a lot of baby fairs, and we butted heads all the time. We did one show in Torrance, California, about 35 miles from where we both lived. It was a terrible show; I sold a total of 27 Little Shirt Anchors the whole weekend. But at least I could add some names to my customer mailing list. At all these shows, I got customers' addresses by having a drawing at my booth; each customer filled out a card to try to win a hot air balloon flight (a friend donated the flights for me to use as a prize). I also let anyone who just stopped at the booth enter the drawing, so I could add them to the list as well.

I had also just gotten a great account with a chain of stores out where I lived, and I mentioned it to a couple of friends at the show. I was really excited about it. I knew that my competitor had also tried to get that same account for some time. But I had the edge: When I was 16 I had worked in one of the stores, so the owner gave my new product a chance.

Two days after the baby fair, I got a call from this new account to come collect all the Little Shirt Anchors they had ordered. When I asked them why, they told me that someone who had purchased two of them from me at a baby fair over the weekend came into the store and really made a scene. The chain didn't expect this kind of trouble when they bought my product, and they certainly didn't want any more.

(continued on next page)

I felt just awful. I couldn't imagine anyone getting so irate over a little thing like a Little Shirt Anchor. I also couldn't imagine someone driving all those miles to return something so inexpensive. Why didn't they just call me directly? My name and phone number were all over the package. I decided to get out the cards from the balloon drawing, call all my customers, and find out who was so mad and why.

I made the calls. No one had any problems. Of the 27 who had purchased Little Shirt Anchors, only two customers had purchased two Little Shirt Anchors; the rest had purchased only one. When I spoke with them, they reported no problems at all.

Now, this was really fishy. I wondered how the person who'd complained to the store even knew it carried Little Shirt Anchors; they'd been in the store for only four days! I drove to the store only to hear all the employees tell me what an embarrassing scene this woman had caused. When I asked them why they gave her a refund at all, since the purchase was made somewhere else, they said she'd been so loud and obnoxious that they'd had no choice.

At that point it became pretty clear to me that this must be the handiwork of my competitor. She couldn't stand the thought of my product being in that store chain. Of course, the store didn't want to hear anything about problems with competitors, they just wanted their *problem (the Little Shirt Anchor) removed.*

As I was leaving the store with my tail between my legs, I noticed a video surveillance camera. I quickly found the manager and said I bet I could describe this obnoxious customer exactly. Could she possibly find the spot on the tape that showed her? The manager reluctantly complied. And there she was: my competitor, in living color, making a fool of herself and the store employees as well.

The manager agreed to keep my product and to require identification in the future before making refunds. A phony plant from my competitor would hesitate to leave a real name and address with the store manager.

The lesson of my story? Be a canny competitor. Generally it's a waste of time and energy to worry about what your competition is up to. But it's not a waste of time to keep yourself informed, to know what your competition is doing.

MAKE SURE YOU HAVE ORDER FORMS TO PASS OUT; MANY PEOPLE WILL ASK FOR THEM. I've gotten orders from my shows up to four years later! Often people don't buy at the show itself, but that doesn't mean the product isn't appealing. Remember that those in attendance had to pay to park, to get in, and to buy food and drink, and by the time they get to your booth they're probably already out $20 to $30, depending on whether they have kids with them. Many consumers attend these shows to browse and collect samples, and giveaways. But they may order something later.

After the drawbacks I've mentioned, you might be wondering why these shows are held at all. Well, some people do make money—generally, exhibitors who have a large number of products and travel along with the show, from city to city. The show's promoters also stand to make a profit if enough people attend, because they charge an admission fee. And large manufacturers often find it worthwhile to attend these shows and hand out samples or coupons for their products, figuring the cost into their advertising budgets.

You really have to do well to come out ahead in a consumer fair, and it doesn't happen very often.

If you do decide to do a consumer show at this early phase of your product's life, figure it as a cost of market testing. Instead of doing the show to earn money, go at it with the thought that you are going to get valuable publicity and one-on-one feedback from your potential customers. Those are the best things a consumer show can offer a vendor with one product.

TRADE SHOWS

Trade shows differ from consumer shows in that the public is not invited. These shows are basically intended for store buyers and the media. You should be able to find several publications at your local library that list such shows by product category so you can pinpoint which would be appropriate for you. Trade publications also have listings and announcements for upcoming shows.

Before you consider buying space at a trade show, you should attend one yourself. You need to see firsthand what goes on and how they are run.

The Scene

In most cases, the major suppliers of your category of product will be there. They will have spent several thousand dollars on their booths and showrooms to make them look inviting. If you think you'd like to become an exhibitor, look at them carefully and gather ideas. You'll have to consider what kind of look you want to use to attract colleagues and customers. A number of companies handle show displays and booth design, but these services usually cost quite a bit; you'll probably start out planning and building your own.

It's really tough to compete with the big companies at these shows, but it's a good way to get your product seen.

You'll notice everyone wears a name badge. Badges do much more than identify participants by name; they identify everyone by title—buyer, sales representative, manufacturer, member of the press, or visitor. I've noticed that people you meet at these shows often look at your badge instead of your face. Don't do that! It makes others feel as though you'll pay attention to them only if the title on their badge warrants it. Sometimes buyers wear a different badge just to get in and out of booths and showrooms without being cornered. If you become an exhibitor, be courteous to all who show an interest in your product.

Getting Involved

If you've visited some trade shows and decided you want to participate, you'll need to send for information far in advance. You'll have to pay for your booth space in advance, design your booth in advance, and arrange for and pay for everything from trash cans to electricity in advance.

As with consumer shows, you must also be prepared to take orders for your product—this time not from consumers but from store buyers. Although many buyers will simply want information about your product or will leave a card so you can mail information to them, some will want to place an order on the spot. You'll want to offer some kind of "show special" to encourage this—some small discount or incentive like prepaid freight.

It's really tough to compete with the big companies at these shows, but it's a good way to get your product seen. Whether it's something you want to do now or to consider for the future is for you to decide, based on your research.

ESTABLISHING RELATIONSHIPS

If you are attending a trade show just to snoop around, or perhaps to locate a potential company to license your product, remember that exhibitors are at the show for one reason and one reason only: to sell *their* product. They have not gone to all the expense of attending in order to listen to you talk about your idea or show them your product.

You want company execs to recognize that you aren't just a pesky inventor thinking of yourself and your product above all else.

So how do you make contact? Have some business cards printed up, walk up to a target company's booth, and introduce yourself as an individual with a product idea you'd like to submit. Ask if the person in charge of your product area is attending the show. If so, simply go up, introduce yourself, and ask to exchange cards. Then say you wouldn't dream of taking up his or her valuable time now, but you will be in touch a few weeks after the show. This will get you off on the best foot with company execs. They'll recognize that you are not just a pesky inventor thinking of yourself and your product above all else; you obviously understand that there is a time and place for everything.

In fact, they may be so impressed that they tell you, "I happen to have a minute or two now if you'd like to discuss your idea." Wouldn't that be something? It's happened to me.

I have also been on the receiving end. As an inventor at heart, I always look at what others have to show me when I'm approached at trade shows. But it's still a waste of time—theirs as well as mine. I'm thinking about the next appointment I have or about who is waiting for my attention in the showroom. The last thing I want to do is review a new product. The details go in one ear and out the other.

More Ways *Not* to Solicit Opinions of Your Product

Store owners are not in business to give you lessons in presenting your idea. Don't waste your time. Don't waste their time.

In my consulting business, many of my eager clients want to take their first sample around to the stores to see if they might order it. Do yourself and the stores a favor: Don't waste your time. Don't waste their time.

Store managers and owners are very busy people. They are not in business to give you lessons in presenting your idea or in pricing, invoicing, or packaging. They may either reject you outright or tell you your product is great—just to get you out of their hair. By being too persistent, you may even create an enemy out of a potential ally. There are many better ways to find out what people think of your idea.

If you're a regular customer at the bait and tackle shop down the street and you want to run your new idea past the owner, that's a different story. There's no pressure—you're not asking someone to sit down and pass judgment, just casually soliciting an opinion. And because you're a customer, the owner will probably be willing to listen and give you some advice. Make a purchase at the same time and you'll definitely stay in the owner's good graces. But remember, this situation is the exception.

If you do get a store owner to offer an opinion on your product, keep in mind that what you're hearing is still just *one person's* opinion. Even a store owner, a potential buyer, can be wrong. Just think how most store buyers probably reacted the first time they saw a Teenage Mutant Ninja Turtle. That didn't stop it from becoming the hottest thing on the market.

Getting By Without a Little Help From Your Friends

Ready for another "don't"? Don't make decisions based on advice from people you know.

It upsets me so much when inventors overrely on the opinions of their friends and relatives. Of course, we all eagerly tell those close to us about our product. It's natural to want to share your great discovery, to get positive reinforcement. But you must remember to take their opinions with a grain of salt. Are the people you are asking any more knowledgeable about inventions than you are? Do they understand the factors involved in getting a product to market? Do they understand manufacturing and distribution? If they do, then their opinions deserve to be given a little weight—but not much.

Tread very carefully when it comes to enthusiastic responses from those you know.

Friends and family do not want to hurt your feelings by expressing a negative opinion. Have you ever asked your friends for their opinion of your new outfit? Your new painting? Your new car? How many times have they come out and said, "Gee, that dress really makes you look fat"? Or "Where in the world did you get that painting? It's so ugly!"?

People you're close to won't provide the unbiased opinions you need—the kind you can get from strangers. I can't emphasize this enough: Tread very carefully when it comes to enthusiastic responses from those you know. Sure, you'll ask for their opinions because you feel compelled to—it's human nature. But don't go off the deep end when they tell you how great your product is and how rich you're going to be. In most cases, they are telling you what you want to hear.

Who Do You Trust?

You should also be wary of praise offered by manufacturers or suppliers for your product. (See more on manufacturing in chapter five.) They have a very good reason for telling you your product is fantastic: If you believe them, they may get some business from you. Same thing with patent attorneys (chapter four). They may be very excited about your new blockbuster idea and encourage you to get protection on it right away so nobody else steals the idea. Who benefits? The attorney, of course, with a fee from you.

Are you beginning to feel like you can't trust anybody? I don't want to make you paranoid, but I don't want you to think everything you hear is true, either. That's the business world. Consider the source and use your best judgment. Of course you want to believe that everyone thinks your idea is the best thing since sliced bread. But you have to realize that many people will say what you want to hear in order to profit off you—without helping you to profit yourself. The person to trust is yourself.

CHAPTER THREE

MONEY MATTERS

You have a great idea, you've done your homework, you know your costs, your markets, how your product will be packaged, where you'll store it, what your overhead will be, how to contact your customers . . . everything except how to pay for it all.

What?

You say you don't know all this?

You'll need to. You'll be asked about all these things by someone who may help finance your product. And in many cases, whether you can find financing will determine whether you are able to go ahead with your project.

You'll find the information you'll need to know and factors you should consider as you read along in this book; but since the money issue is probably looming large in your mind, I'll tackle it right off the bat.

WHERE TO START

When you're ready to look for funding, the first place to look is to yourself. After all, who has more belief in your product than you? It's a pretty easy sale. But don't even think about planning to spend every penny you've got and go into debt up to your eyeballs to finance your idea.

I once had a client who got a $100,000 mortgage on his home in order to finance his product idea. Unfortunately, he didn't ask for my advice beforehand; I would have tied his hands behind his back so he couldn't sign anything.

That is an awful lot of money. Most products simply can't support that kind of investment. When I hear about someone spending such money, it tells me they "want it all" now. They want to start big and skip the learning curve.

Most of the time, this is not the best approach. What is? Starting as small as you can and working your way up; doing things in steps and paying for as much as you can yourself.

The Business Plan

Even if you plan to get your funding from Aunt Peggy, you should sit down and create a business plan—both for yourself and for your generous aunt. Don't worry: You don't need to be an accountant to do it.

I have included a sample business plan in the Library of Resources. Take a look at it if you're not familiar with them.

First, list your one-time expenses. Maybe you need to buy a fax machine, or tooling for your product. You have to get a business license, pay warehouse and utility deposits . . . list any costs you can think of that only have to be paid once.

Next, list all the costs you'll have on an ongoing basis, aside from those directly associated with your product or inventory. These might include phone bills, warehouse rent, automobile expenses, and office supplies. You'll need to do estimates on many of these things; just give it your best shot.

Now list all the costs associated with the product itself. Don't forget to include packaging and shipping charges. You also need to include labor costs, even if you are doing part of the labor yourself. You will eventually have to get someone else to do it, so you'll want to estimate that cost here. (Chapters five and seven cover the costs of manufacturing and packaging your product.) Be aware that your product costs will go down as quantities go up. Keep all your product costs separate from your other costs.

Don't worry: You don't need to be an accountant to draw up a business plan.

You'll end up with three cost totals: start-up or one-time expenses, your ongoing monthly expenses, and the cost of your product. From this you will be able to get an idea of how much money you'll need now to get started, as well as how much you'll need the first year to keep going.

A caveat: I have reviewed dozens of business plans, ranging from a few thousand dollars to several million. I have never, ever seen a business plan that came close to predicting what really happened in a business. It is impossible, without a crystal ball, to predict the future. But that's essentially what a business plan tries to do. The important thing is not to get every figure exactly right; it's sitting down and thinking about what you need to do—and showing whoever you go to for financing that you've done that thinking.

I have included a sample business plan in the Library of Resources that my partner Nancy and I did to enter a grant program. This is really a summary business plan, as it is very brief but does state objectives, goals, sales strategies, and estimates on expenses and income. A complete business plan would require more detail to support how you arrived at all your figures.

Sales Projections

Now it's time to estimate sales: This is when you really need to be a fortune teller. Should you throw a dart at a board? Pick a number out of your hat? Even the biggest companies, with mountains of sales data to study for their particular industry, have a very hard time doing sales projections. It requires much skill and much luck.

Estimating sales requires much skill and luck. The key is to gather as much information as possible.

The key when estimating sales is to gather as much information as possible and do a prediction based on logical thinking. You can assume that a fraction of a percent of the potential customers available will buy your product, due to economics and product availability. (For example: Do they have the money available when they see your product? Did they see your product in an ad but can't find it in a store?)

You may wish to base your estimates on how many retail outlets you could sell to. Trade organizations, such as those that put on the trade shows, should be able to give you information on the number of such stores in the country. I'd advise sticking to specialty stores for your projections. Skip the big guys for now—they're a long shot at best for a new product and a new manufacturer. Figure them into your plans for year two or three.

Whatever method you use to determine your projected sales, write down how you came up with the figures. And include three sets of figures: a low estimate, a moderate estimate, and a high estimate. By doing this you'll be covering all the possible scenarios and you and your investors will have a more realistic expectation of the outcome.

FINDING FINANCING

As you launch your own business, you should be aware that all businesses follow certain patterns. For example, most new ventures will start to show a profit after three to five years. The longer you are in business, the better your chance of success. If you make it through the first year, you're looking very good. The number one reason a new business folds is money—specifically, lack of money, or undercapitalization.

The number one reason a new business folds is money—specifically, lack of money.

Be Prepared

This doesn't mean you must start out with $50,000 on day one. You may only need $1,500 to get started. But you should expect the unexpected. When an opportunity comes up, you don't want to miss out because you couldn't come up with the financing to back it. I'll give you a couple of examples.

Before I had any business experience, I was convinced that my Little Shirt Anchor needed a great package, one that would both show people how the product worked and make them feel good about buying it. I designed a package that had a photograph of a baby wearing a Little Shirt Anchor and playing with Dad outdoors on a nice sunny day. It was designed to hang on a pegboard, where you might find pacifiers and teething rings.

I began to place the product in some small baby specialty stores. I also attended baby fairs just about every weekend, renting a booth space to gain exposure to retailers and the public. A pattern soon emerged: In the cities where I attended a baby fair, stores said that my product did very well. In other areas, stores complained that sales were sluggish. I was dangerously close to losing some of those accounts, yet I knew from baby fairs that the public liked the product. Parents purchased Little Shirt Anchors readily after seeing them demonstrated. So I concluded the packaging was somehow the weak link and decided to develop a new package.

I hired a designer this time—a considerable expense, but one I considered worthwhile. I wanted to get it right this time. In addition to redesigning the package, I also asked him to design a point-of-purchase display (freestanding or designed to sit on store counters; see chapter eight). This whole learning experience cost me just under $3,000. The new package was terrific, and the point-of-purchase displays really helped sales take off.

You can't know in advance what will or won't work, and mistakes, while instructive, can be costly.

Moral, part one: You can't know in advance what will or won't work, and mistakes, while instructive, can be costly. So make sure you have a mistake cushion.

On to my second example:

After I had been in business for about a year and a half, I became a vendor for JCPenney. I had been working on getting the account for a while, but I had not been in contact with the store for about five months when I received a call from one of their buyers, who told me Penney's would like to carry Little Shirt Anchors. After a quick conversation, I agreed to send her samples for a photography shoot scheduled in two weeks.

I was ready to pass out from excitement—people dream of getting their product into Penney's, one of the largest and most respected retailers in the country. I immediately went to my calculator to figure out exactly how rich I would be.

My family and I had four days of jubilation; then came the phone call. My buyer from Penney's told me the higher-ups wanted a different type of package for the Little Shirt Anchor—a blister package. And they would still need samples of the Little Shirt

Anchor, in the blister package, in time for the photo shoot. Would that be a problem? "Of course not," I answered. "No problem."

I hung up the phone and dashed to my *Yellow Pages* to find a packaging company. The first order of business was to find out what in the world a blister package was, and then find out if I could possibly get a sample made and delivered to the Penney's buyer before my deadline.

I found out that a blister package is similar to the type of package you might buy batteries in: A plastic bubble holds the item, which is then sealed onto a card. Oh, the card is not just any ordinary card, either. It must be coated with a special material so that when it is exposed to heat, the blister will adhere to it.

The smallest order the packaging company would accept was 25,000. The cost? A mere—hah—$3,800. And Penney's needed the new package when? In 14 days! Was I still happy Penney's had called? Of course; all this was just the price of doing business. But if I hadn't been able to lay my hands on $3,800 in a hurry, the account would have been lost. Instead, I received my first order from Penney's nine months later.

Moral, part two: Expect unexpected expenses and be prepared for them.

Start-Up Money

How to get money? Well, if you are married and working with your spouse, you may want to consider a mortgage loan. Or you may have family and friends who would like to invest in your business. You may find that people you know *offer* you money for your venture before you have a chance to ask. If this is the case, you need to decide whom you're comfortable borrowing money from. I know I would never want to owe anything to certain members of my family—even if it was the only alternative! You will have to live with this decision a long time, so choose your investors wisely.

And make sure you don't cut yourself short. Remember, you came up with the idea for the invention, and your sweat and sacrifice will be responsible for its success or failure. Letting others ride on your wagon will give you the capital to move forward—but you don't want to wind up having them own your company.

Investors and Lenders

A CPA gave me a simple formula for deciding how much share in a company an investor should get. Imagine that you suddenly need to sell your company. What amount would you accept for an outright buy? Let's say $100,000. Let's also assume your Aunt Shirley wants to invest $5,000. Then your Aunt Shirley would be entitled to 5 percent of the profits from your company. Of course this is just a rough gauge—and you may choose not to use it at all.

Be very clear about the arrangement when your investors come on board. Communicating now will prevent problems in the future.

The best advice I can give you is to be very clear about the arrangement when your investors come on board. Don't say "I'll take care of you" and assume your investors know what you mean. You may feel generous offering a 3 to 4 percent share. They may feel cheated with anything less than 12 percent. Communicating now will prevent problems in the future.

The biggest problem with borrowing from friends or family is that it is usually done on a casual basis, with no regimen for getting paid back or for equity in the company. It's very important to put everything in writing at the time of the loan. Don't make an open-ended deal that you'll have to hash out at some later date. If you're going to give investors part of the company in exchange for the money, for example, you need to determine how much of a part. This is difficult for an inventor to do, because your idea is worth nothing without the money to put it out there. And investors need to be aware that they may not get any money back if the product does not do well.

A straight loan can simplify things. You don't have to worry about feeling guilty as you might if you agree to give an investor a portion of the profits, then your product fails. But you will still have to evaluate the situation carefully and get everything in writing. For a straight loan, you should draw up a contract with a repayment plan already in place. If interest will be charged, determine how much for how long. Your lender should also consult an accountant about the tax consequences of lending you money.

Bank Loans

If you want to try to get money from a bank based on your product idea, think again. A conventional bank loan will require a guarantee and, most often, collateral—perhaps a mortgage on your home or maybe on your boat or RV, if you have one. Banks are not in the venture capital business and are not going to start with you.

Even if you show a loan officer a purchase order from Sears for 50,000 units of your product, the bank will still be reluctant to give you a loan. Remember, there is a big difference between getting an order—complete with delivery date and cancel date—and filling that order. Many things can go wrong in the interim, and banks know it.

Other Options

Here are a variety of other sources to investigate for funding:

The Small Business Administration. The SBA guarantees loans through local banks, but such loans are usually only given to existing businesses with a proven track record and a well-written business plan. You may have to get some help to succeed in this venture. And figure you will spend about two months on all the paperwork—a real mountain of forms and documents.

Small business development agencies. Your bank and the SBA can help you locate them. This may be your best shot at financing; these agencies are sometimes willing to loan small amounts of money to individuals or to help you pay certain costs associated with your business. If you are a woman, or if a woman owns 51 percent or more of your business, you will usually qualify for minority business owner status, which can help you secure a loan.

Grants. Essentially money you don't have to pay back, grants are available through the federal government as well as state and regional outlets. They are much harder to obtain than loans, but it's well worth the time and effort to find them and apply.

Check your local library or the small business development listing in your telephone book to get started on the right track. And when you call for information, don't give up too easily. The person you speak to may not be able to assist you but may know someone who can. Be sure to ask for referrals and for help.

Venture capital monies. A venture capitalist is an individual or company that invests in "risky" ventures that may pay off big. Venture capitalists do not require collateral; they do not require that the money be paid back. They do, however, require a pretty sizable chunk of your business in return for financing.

One of the benefits of dealing with a venture capitalist is that you have a silent partner (or maybe not so silent, depending on the deal), who is a valuable resource because he or she will make sure you get the advice and help you need to run your business. A venture capitalist wants to see his or her investment grow as much as you do.

Where do you find someone who likes to give away money? Try asking professionals you work with—bankers, attorneys, accountants—if they've dealt with anyone who might be interested.

Proceed cautiously when dealing with venture capitalists. Keep in mind that they have done this sort of thing many times before, and you have not. Have your attorney examine any deal you agree to and help you decide if it's worth it. If you don't have an attorney, this is the time to get one.

CHAPTER FOUR

INVENTOR'S PARANOIA AND HOW TO AVOID IT

Protecting Your Product Idea

Have you been feeling a bit nervous and protective about your great idea ever since you dreamed it up? Have you felt a slight twinge of worry even when telling your closest friends about your new product? Don't feel guilty—your feelings are perfectly natural. You've got a simple case of inventor's paranoia, and there are several ways to treat it.

Protecting Your Idea as an Individual Manufacturer

The possibility of someone stealing your idea is a very real concern, but probably not in the context you imagine. It's very rare for a product to be knocked off until it becomes quite successful.

The first copy of your product may come from someone like yourself—an individual manufacturer who has either come up with the same idea independently or has developed another version of your product. This is not someone you need to be concerned about.

No, I'm not crazy!

When two small manufacturers make the same product, what usually happens is both companies do well. Store buyers probably won't distinguish one product from the other, unless their prices are drastically different. If they're successful with one, they'll be successful with the other.

When two small manufacturers make the same product, what usually happens is that both companies do well.

Small-Potatoes Competition

Let's say your competitor has been to see a buyer five times, and each time the buyer turns the product down. The next visit may come from you—and you may get the sale based on the other company's persistence. Or vice versa. If you get a write-up in your local paper, your competitor may reap the benefits of your publicity.

I know it's hard to imagine that your product could be confused with another product, but it happens all the time. Think about the last time you saw an ad for a particular item you were interested in. You probably remember the general concept rather than the name of the company that manufactures the product. If you want to buy it, you may call your local drugstore and say, "Do you carry that new automatic page turner for books? I read about it in the paper." If the clerk says, "We just got some of those in," you'll likely hop down to the store and buy one—whether or not it's the same brand you saw advertised.

Stores and consumers are not particularly discerning. Only inventors think their product is vastly superior to anything similar on the market. Not that your product *isn't* superior—but it's the concept, not the features, that is going to sell it in the beginning. Now, looking for a new refrigerator or a car is a different story. This is the realm of comparison shopping. But for the new, unique, never-heard-of-before product, it's the big picture that counts, not the details.

It's hard to imagine that your product could be confused with another product, but it happens all the time.

Generally, at this level, competitors are nothing to worry about; they'll boost public awareness of your product as well as theirs. And there are plenty of customers to go around.

Big-Time Competition

The knockoff artists you need to be worried about are the big companies—the ones whose brand name makes a difference. They have established relationships with buyers at stores, and chances are the stores already buy other products from them. So when they offer a copycat product to retailers and say, "We've got a new product in our line that turns the pages of a book automatically, and we're going to advertise it all over the country, and we'll even mention your store name in the ads," that's when you don't have much of a chance to compete.

But there is good news. Big companies have to spend several thousand dollars on a new product. They're not going to do this unless they're pretty sure it's going to sell in big numbers. The likelihood of getting knocked off by a big company in the beginning is very, very low indeed. It's when you become a success that the big guy will want your idea.

Confidentiality Agreements

You can take steps to protect your idea early on, while undertaking your initial research and evaluation, as well as later, when you are manufacturing it or presenting it to companies to get it licensed. The first type of protection you'll want to get is called a *confidentiality agreement*.

A confidentiality agreement shouldn't be viewed as an indication of distrust; it's simply the proper way to conduct business.

A confidentiality agreement, as the name implies, states that the signer will not share your idea with anyone else. Before you reveal your idea to anyone—say, a manufacturer who's going to give you price quotes or an illustrator who's going to prepare artwork for your package—it's wise to ask them to sign on the dotted line.

Don't be embarrassed about asking someone to sign an agreement. It is standard procedure when dealing with proprietary ideas and shouldn't be viewed as an indication of distrust; it's simply the proper way to conduct business.

You will need to see an attorney to get a confidentiality agreement for your own use, but I've included an example of one in the Library of Resources. I've used an agreement like it in the past.

Keeping Records

It's very important to keep organized files and notes on all matters relating to your product idea. You should have:

- all signed confidentiality agreements
- all signed non-confidentiality agreements
- all submission documents
- all correspondence received and sent
- a detailed phone log, with dates, times, subjects discussed, and a copy of your phone bill
- records of samples sent (a photo is best)
- all receipts for shipping or postage used to send literature or samples
- notes on how you located the individual or company, for example, whether you found them listed in a phone directory, library listing, advertisement (where and when did it run), through a referral, etc.

What If They Won't Sign?

If someone refuses to sign a confidentiality agreement, ask what he or she objects to. If you can't get their cooperation, you'll have to decide whether the benefits of disclosing your idea outweigh the risks. An attorney can help you with this decision.

Be aware that it's common practice for companies to refuse to sign a confidentiality agreement when you submit a product for licensing consideration (a process covered in chapter six). It would be impossible for a company to review an idea without disclosing it to sales reps, manufacturers, or other employees; you may be told that signing such an agreement would go against company policy. Again, you'll have to decide whether it's worthwhile to share your unprotected idea; and again, I advise you to consult an attorney.

Note: Be sure you don't confuse a confidentiality agreement and a *non-confidentiality agreement.* The latter is generally used by a company when it is considering whether to license an outside idea and states—again, as the name implies—that the company will *not* keep the idea in confidence.

THE NEXT STEP: PATENTING YOUR PRODUCT

When you get beyond the idea stage and start to manufacture and sell your product, you'll want to consider taking further steps to protect it. The U.S. government offers several types of protection for new products:

- A *trademark* is a word, name, symbol, or device (or any combination of these) used by manufacturers or merchants to identify their goods and distinguish them from those made or sold by others.
- A *copyright* protects the creative work of composers, writers, artists, filmmakers, and others; a copyright's term is currently life plus 50 years and is renewable.
- A *patent* gives an inventor exclusive legal right to exclude anyone else from manufacturing, selling, marketing, importing, or

using an invention during the life of the patent. Patent life is currently 17 years and is not renewable.

As the creator of a new product, you will most likely be concerned with the third type of protection—a patent.

When you think about patenting your product, a lot depends on whether you will be manufacturing it yourself or licensing it to a company to manufacture. In the former case it's really up to you to decide how much protection you need for your product. In the latter instance, your decision will probably be made for you by the company that licenses your product.

I would hate to see anyone not pursue a product idea because they don't have the money for a patent; on the other hand. . . .

Should You or Shouldn't You?

Probably the best thing a patent can do is act as a deterrent to someone who considers knocking you off. A patent does not actually prevent someone from copying your idea. It's like putting bars up on the windows of your home. It is certainly going to put off most burglars, but a determined burglar will just climb down your chimney, soot and all.

So is it worthwhile? And if so, at what stage? The best person to answer those questions is a patent attorney (more on this topic below).

A patent is a large expense for an inventor. A *utility patent*, which, in layman's terms, covers a new idea or concept, usually starts at about $2,000 and goes up from there, depending on the details of the patent application. A *design patent*, which covers a particular design of an existing product—say, a toaster with slots on the bottom instead of the top—can run anywhere from $800 to $1,800, in my guesstimation. Many people apply for both types of patents. Again, you'll need to consult with a patent attorney for advice on whether you should do this.

The average patent infringement case will cost the patent holder (you) $200,000; and you have only a 50/50 chance of winning.

I would hate to see anyone not pursue a product idea because they don't have the money for a patent; on the other hand, I hate to hear that a product idea has been ripped off and the inventor has been left with no recourse.

APPLYING FOR A PATENT

You have one year from the date of your first public sale to file for a patent; after that, you lose your right to one. To get a patent on a product, you must file a *patent application* with the patent office in Washington, DC. Completing a patent application is not like filling out a charge card application or an employment application. It requires very precise details. You must supply drawings that indicate the precise dimensions of your invention. You must even label your product's features in a specific type size.

You have one year from the date of your first public sale to file for a patent; after that, you lose your right to one.

In your application, you need to address the claims you are making for your product idea and note specific variations between your idea and any similar patents. I've included a copy of my patent for the Little Shirt Anchor in the Library of Resources to give you an idea of what you'll get.

Patent Searches

A *patent search* is sometimes done in advance of a patent application, to determine whether other patents have been issued that may conflict with your product idea. As I noted earlier, the fact that you've never seen a product like yours before is no guarantee it doesn't exist—or didn't exist at one time. Patents date back to the early 1800s; you may be infringing on a patent by Abraham Lincoln!

You can do a patent search at large public libraries. Many have indexes and directories, as well as complete patents on microfilm. The reference librarian will be able to help you. Even a thorough search cannot guarantee anything, however, because patent records are routinely updated every three or four months and the source you checked may be out of date.

Getting Help: The Patent Attorney

Several books on the market offer instructions on filing a patent, but I've never used one and so can't vouch for them. Some of my clients who have tried them, however, have met with less than total success.

Your best bet is to find a good *patent attorney*. A little metaphor will help explain my reasoning: Imagine I gave you a book that provided you with complete instructions for driving an Indy race car, then told you to race the experts. How successful do you think you would be? Sure, you'd know the mechanics—ignition, brakes, gears. . . . But a certain amount of finesse and experience are required to compete properly and not get hurt.

> **Why risk making a mistake and potentially losing your rights? Hiring a patent attorney is well worth the money.**

It's the same with patents. You can read the application and do your best at applying. But why risk making a mistake and potentially losing your rights? This happened to a consulting client of mine: She kept getting her application back, with notes pointing out her errors, and, despite a helpful patent examiner, she just couldn't get it right. After passing her one-year patent window, she lost her patent rights.

Patent attorneys are not inexpensive. But I feel they're well worth the money.

What to Look For

A good patent attorney is someone who takes your calls and doesn't rush you off the phone. Someone who doesn't mind answering "dumb" questions. Someone who will occasionally tell you that you don't need to spend more money. Someone who knows what's what.

I'm fortunate to have the best patent attorney in the world. At least I think he is. He does all of the above, and he's been a valuable source of information and help to me. I often refer my consulting clients to him.

Where to Find One

How did I dig up this gem? I began my search by asking around, but the attorneys I was referred to just didn't feel "right" for me. So I opened my phone book to "Attorneys; Patent, Trademark, and Copyright" and began dialing.

Ask lots of questions. I armed myself with a list of questions, and you should too: What do you charge for a design patent application? How long have you been practicing as a patent attorney? That sort of thing. When I called each number, I asked to speak with the attorney; if he or she wouldn't come to the phone and I was told I had to come in for a consultation, that name got crossed off my list. Ditto if the attorney came to the phone but was reluctant to answer my few simple questions.

Keep in mind that you are the potential employer; the lawyer is the potential employee.

Be choosy. When you call around, keep in mind that you are the potential employer; the lawyer is the potential employee. If you're asked to pay for the initial appointment, politely decline and move on. Of course, you should make it clear that you only need 10 or 15 minutes; after all, lawyers make a good part of their salaries by billing for their time, so it's understandable that they wouldn't want to give you a lot of their time without charge.

Get referrals. You can also call the American Bar Association for a referral. Every attorney must pass the State Bar Exam before practicing law. The Bar Association in your state has a listing of attorneys. Be aware that only registered patent attorneys can practice law before the Patent and Trademark Office or represent you in court; when you call, make sure they have this credential.

Mr. or Ms. Right

After many disappointments, my great attorney search finally turned up someone who not only talked to me but genuinely sounded happy to answer my questions. He explained how he could help me, and when I asked him about prices he gave me estimates but told me he'd have to see my invention before he could be more precise. After at least 30 minutes on the phone with him, no charge, I felt comfortable hiring him, and he has worked on my patents for the Little Shirt Anchors and subsequent products. He reviews my submission agreements and checks out all my contracts before I sign them.

My patent attorney has been a huge help to me; if I could clone him I would. His counterpart is what you're looking for. Take the time to shop around and ask the right questions—and follow your instincts.

No company is going to pay you a royalty on an idea that is not protected.

PATENTS AND LICENSING

If you are thinking about licensing your product idea (see chapter six for more on this subject), you will almost certainly want to talk to a good attorney about getting some kind of patent application on file before you officially present your idea to anyone. The reason is not to protect your idea from being ripped off, although that's a nice side benefit; the real reason is that no company is going to pay you a royalty on an idea that is not protected.

The Worst-Case Scenario

Let's say you sign a license agreement with Black & Decker without first patenting your product. The company begins the costly process of researching your idea, how it should be packaged, where it should be made, what price point to strive for, etc. One or two or three years later, the company finally gets your product to market, and it pays you a royalty every time one is sold. Very nice. But wait—Black

& Decker's competitors have copied your product, and since they don't have to pay you a royalty, they are able to sell their version more cheaply than Black & Decker can. Now both you and your company are left out in the cold, neither one of you happy with the arrangement.

In this case, you can see the importance of having a patent. This is a simplistic example, and it doesn't apply to every product; once again, I urge you to seek the advice of a competent patent attorney.

PATENT PENDING

Patent applications are reviewed by an examiner at the patent office, whose job is to make absolutely sure your patent does not conflict with any existing patents. Often, the patent office will return a patent application to the submitter, pointing out specific information that must be supplied before resubmission. This is called rejection, but you shouldn't take it personally—it's quite common.

Don't take a patent rejection personally—it's quite common.

Once you have submitted a patent application, your product idea has *Patent Pending status*. You've probably seen this designation on packages and labels; it means the patent application is in the review stage at the patent office. Patents that are pending are not available for public scrutiny, in order to keep other individuals from knowing what claims you are making on your application. This is valuable information and must be maintained in secrecy until the patent has been issued.

I received a *utility* patent issued for my Little Shirt Anchor within nine months of my application; that's record-breaking time—normally it's a very lengthy process. On the other hand, I've been waiting for a *design* patent on another product for three years!

After you get a patent, however, it's good for 17 years from the date of issue. Then, although you've still got the patent, the idea behind your product becomes *public domain.* Anyone can use it, without your permission, though the idea cannot be patented again by someone else.

OTHER WAYS TO PROTECT YOUR IDEA

You should always keep diligent notes on the development of your product. Start with what made you think of it, then record when and how you built and tested prototypes, and whether and when you made changes to the idea. If you discover similar products on the market or previously on the market, write that down too. Keep your notes in a safe place.

If you contact a company or individual by telephone to discuss your idea, make a copy of your phone bill with notes on the call and what was discussed. Keep copies of both the bill and the notes with your journal. If there is ever a dispute in the future as to who came up with the original idea, when the idea was conceived, who was exposed to the idea, in what capacity and to what degree—or who

More on Patents and Trademarks

For more information on patents and trademarks, write to the Commissioner of Patents and Trademarks, Washington, DC 20231, or the U.S. Small Business Administration Office of Innovation, Research and Technology, SBIR, 1441 L St. NW, Washington, DC 20416.

The patent office has a telephone message system which you may access for general information on procedures for filing patents, trademarks, or copyrights with explanations on the differences between each protection service and how they may apply to you.

Each of these offices can provide you with literature that will give you an overview of the protection process or a detailed summary. Most of the brochures and booklets are written so that a layperson can understand them.

worked on the idea—you will have very impressive accountability. This may be necessary for legal disputes.

Mail Safe

If you send samples through the mail, or paperwork of any kind describing your idea, use certified mail, Federal Express, UPS, or another method that provides confirmation of delivery. Keep this confirmation as documentation, along with concise notes on what was sent. It's also a good idea to follow up with a phone call to make sure your mail was received. Make notes about the nature of the call, what was said, and by whom. Keep a copy of the phone bill.

When I send out samples, I take a picture of what I'm sending and attach the picture to a note indicating the date and the name of the person the samples are going to, as well as a copy of any correspondence I'm sending with the samples. If you have several different prototypes or different versions of your product, this will help you keep track of exactly what was sent and whether all the components are returned.

Make sure you don't leave samples floating around; do your best to retrieve them once they've been examined. Samples that sit around in someone else's office may be seen by all sorts of unintended audiences.

> **Make sure you don't leave samples floating around. Keep track of what was sent to whom, when, and why.**

IF YOU'RE RIPPED OFF ANYWAY

You may send your product to the most reputable company in the world, get signatures on 15 agreements stating that the company will not steal your idea, and then find that one of the secretaries has left to start her own business—with your product! The truth—and you might as well accept it right now—is that inventors are always faced with the possibility that someone will rip you off.

As I wrote this book, I was aware of two other companies making a product similar to the Little Shirt Anchor, designed to keep babies'

At some point you have to decide you've done all you can to protect your product and jump in and get to work.

shirts tucked in—direct infringements on my patent. Neither was a major company; in fact, I will be surprised if they're still around a year from now.

How did I feel when I found out about these copycats? Well, the same way I feel about individual inventors with ideas similar to mine: There are enough babies to go around.

If you're so afraid of being knocked off that you never show your idea to anyone, you will never get it to market. At some point you have to decide that you've done all you can to protect your product and jump in and get to work. I realize it's a scary thing. And there *are* unscrupulous individuals who may decide your idea is worth stealing. But if you get good advice, take advantage of the types of protection that are available, and always keep careful records (and let everyone know you are doing so), the odds of having your idea all to yourself are in your favor.

CHAPTER FIVE

CAN YOU MAKE MY GIZMO?

Pricing and Producing Your Product

Whether you are planning to sell your product yourself—*or license it to someone else to sell*—you will need to do your homework on product pricing and production.

It's obvious that such research is necessary, you say, if you'll be arranging for manufacturing and then selling it yourself. But why, you may ask, do you need to do all this work if you're going to license it to someone else? Won't the licenser be doing the pricing and production?

Ultimately, yes, it will. But when you are trying to interest a company in licensing your product, you need to sell the company on the idea of going all the way with *your* idea instead of the many, many others they have to choose from. And knowing how your product may best be priced and produced is all part of a winning licensing presentation.

Doing Your Homework

Let me give you an example of the importance of knowing all about your product when trying to license it to another company: One inventor I know recently submitted an idea to a major company, only to be told the product would be too expensive to manufacture.

The extra time you spend doing research and gathering facts can make all the difference.

What could she say in response?

Nothing. She hadn't done any investigation on her own, so she had no choice but to accept the rejection.

Being prepared—doing that sort of research in advance—has gotten me past on-the-spot rejection and on to the next level of consideration. It doesn't happen without a little effort, but the extra time you spend doing your research and gathering facts can make all the difference. And every time you present your product, you'll be confident you're giving it your best shot.

Here's an example of the benefits of doing your homework: I once created a game that used magnets and submitted it to a company for licensing. The company loved the game but rejected the submission because the magnets were too expensive.

Fortunately, I had done my homework. I was able to give the reviewer my best magnet price on the spot, a price that was less than half of what the company had come up with. That put my product back in the running. Just a few phone calls' worth of work had prepared me.

Many of my consulting clients and other inventors I have spoken to thought the hard part of their job ended when they came up with their product idea. "Now I want someone else to do all the work and make me rich" is essentially what they said. Well, it doesn't work that way. The vast majority of inventions are not get-rich-quick schemes. What pays is having faith in your idea, patiently taking the necessary steps to reach your goals, and maintaining realistic expectations.

PRICING YOUR PRODUCT

Determining your product's cost is one of the steps that will get you closer to your goal, whether it is to license your product or sell it yourself at a profit. You will need to know all your costs in order to create a business plan (see chapter three).

A product has to be more than exciting and unique. It also has to be affordable.

Why Price Matters

Price is an all-important factor in the success or failure of your product. Exciting and unique as your product may be, it also has to be affordable.

Consider the stapler: a very useful invention that can be found in just about every office and home. What do you think would happen if a stapler cost $69.99? Its marketability would be somewhat limited. What if it cost $295? Many people would find another way to hold papers together.

When customers decide whether to purchase your product, they'll ask themselves whether the quality and usefulness of your product are fairly reflected in the price tag; if not, they'll turn elsewhere. In the case of the $295 stapler, the usefulness of the product is not equal to the cost. Obvious, right? But when *you* are the proud inventor of a product, it's very difficult to see it objectively and judge its value. It's almost like trying to be an unbiased judge of your child. Pricing your product, however, is by no means a purely emotional decision. There are plenty of cold, hard facts to consider.

CALCULATING COSTS

Here's a simple formula to work with:

cost + freight costs* + packaging costs = wholesale price

wholesale price × 2 = retail price

Let's use a real example. You are going to make gloves. Your costs for a pair are:

fabric	$2.40
buttons	.50
cutting and sewing	2.00
freight on material	.14
freight on buttons	.04
freight from sewer to you	.16
plastic bag for package	.05
header for package	.11
labor for packaging	.05

Your cost for material, buttons, cutting, and sewing = $4.90
freight and packaging cost (incuding labor) = $.55 per pair

$4.90 + .55 = $5.45 (total cost).

$5.45 × 2 = wholesale price of $10.90

This doubling of cost represents your gross profit margin on the sale of your product. This must cover all your overhead and leave you with a profit. Please evaluate your individual circumstances to determine if this formula will be best for you.

The wholesale price of $10.90 is the price stores will pay for your product. You can assume that the stores will double, or *keystone*, whatever they pay for the item, which helps you arrive at your retail price of: $10.90 × 2 = $23.80.

---

* Freight costs in this instance are the costs you pay in order to get the materials to you.

Consider Your Options

Now you need to consider whether the market will bear a pair of gloves at $23.80, or $24. If this seems to be a reasonable figure, then all systems are go. But let's say you know for a fact that no glove has ever been sold for more than $18 a pair. That means you have to get your retail price down by almost one-third.

You may have to find a cheaper source for your fabric or use different fabric. You may need to change the style so the gloves use less fabric. Perhaps you should use a cheaper button or no button at all. Would a snap be cheaper? Would a bow be cheaper? Maybe you can find a supplier closer to you to cut down on your freight costs. Maybe you could ask your sewing contractor for advice on bringing the cost down.

Don't Underestimate

If your product is too expensive, now is the time to do something about it.

There are other factors to consider as well. Maybe you packaged the gloves yourself, so you decided not to add in a packaging charge. Maybe you pick up the buttons and fabric so there is no freight charge. But you would be making a terrible mistake by not figuring in those charges, even if you do not pay them directly now. If your product is successful, or if someone else will be making and selling it for you, these tasks will eventually be added in. Added cost means less profit or no profit.

Another way to miscalculate: Maybe the buttons for your gloves cost 25 cents a pair when you make 1,000 at a time, but when you make a million at a time they will only cost 5 cents. So, of course, you base your calculations on the latter figure. But you shouldn't. Always base your calculations on the higher figure. You can always adjust your price downward later if your costs allow.

The Bottom Line

The bottom line in pricing is to sell the product for as much as you can get for it. Make sure you're charging enough to make a profit. If you use the formula we started with and you are within your target, you can feel pretty good. If you are not within target at this point, go back to the drawing board. Get some professional help, if you need it, to help determine your financial goals.

(See chapter seven for pricing information when working with distributors. There are some real differences.)

The point is, you have many options at this stage of development. If your product is too expensive, now is the time to do something about it. Otherwise, if you go ahead and present your product to a company for licensing, the price may be a major obstacle. The company may well do its own research and determine that the product is not cost-effective.

You don't have to be an expert on manufacturing; just find people who can tell you what you need to know.

MANUFACTURING COSTS

Whatever your future plans may be—whether you plan to launch your own business or license your product out to a company for those "big" royalty checks—you need to determine manufacturing cost before you begin. Manufacturing cost is a crucial factor in determining the marketability of your product.

As you prepare to present your product idea to companies, make manufacturing costs your number one priority. Companies evaluating your product will. As I mentioned above, one of the first questions out of buyers' mouths will be, "How much?"

Where to Begin

The first thing to keep in mind: When you are working in an area you are not familiar with, don't be afraid to ask questions! Of course, as a new inventor, you're not likely to be an expert on manufacturing processes; the key is finding people who can tell you what you need to know.

Say your product is made of plastic. There are several methods for manufacturing plastic goods. You need to determine which is best suited to your needs: injection molding, which is used to make plastic bowls and such; blow molding, used for hollow plastic products such as baby bottles; or maybe plastic extrusion, used to make windshield wiper blades and other such products. If plastic is your medium, you should familiarize yourself with these various processes and the manufacturers who use the process most appropriate for your product.

The same is true for electrical items, metal parts, sewn goods, etc. Manufacturers are experts on the materials they work with and can offer all sorts of enlightening information.

Finding Help

An excellent resource for finding manufacturing and materials nationwide is the *Thomas Register of American Manufacturers*. You can find just about anything within its several large volumes of cross-referenced information—doll eyes, metal hooks, foam padding, chicken wire, you name it. Most libraries have a *Thomas Register*, and a reference librarian can help you figure out how to use it. Since it costs about $240 per set, the library's probably your best bet unless you're planning to get a *lot* of use out of it.

The second reference is your local *Yellow Pages* or business-to-business telephone directory. The latter lists companies whose products or services can be of use to other companies, rather than to consumers. (Your telephone company can tell you how to get a business-to-business directory for your area.)

Look under the appropriate manufacturing heading in your directory—"Plastics," for instance. If there's a manufacturer near you, call to make an appointment to discuss the type of manufacturing you need. If you can't find a manufacturer in your area, call the one closest to you and describe your product over the phone—or describe a similar product if you are wary of revealing your product idea.

You may wish to ask the manufacturer to sign a confidentiality agreement, which is discussed in chapter four. Based on my own experience, however, I'd say manufacturers are not in business to steal ideas from potential customers. They don't have distribution outlets and sales reps and all the other resources necessary to successfully market a product. They will be happy to let you take care of all those things while they make their money manufacturing the product for you.

Listen to *Everyone*

You may have to talk to several manufacturers before finding someone willing to spend a few minutes on you and your product. Don't give up! This is a critical stage.

That's right: A complete stranger may make a suggestion that can improve your product.

A manufacturer may tell you the design you have chosen is simply not workable. Or there's another, less expensive way to do it—or a way that may not be cheaper but will make your product work better.

That's right, I just suggested that a complete stranger may make a suggestion that can improve your product. And often this person will be a manufacturer. After all, manufacturers know their materials better than you do.

It's difficult to hear that *your* way is not the best and the only way to make your product. I remember how shocked I was when I first saw a competitor's product based on the same idea as my Little Shirt Anchor, but with a different design. Up until that moment I *knew* there was only one way to make that device. Seeing her design was a real eye-opener.

The point is, you shouldn't be so in love with your product design that you discount what could be valuable advice from others. When you talk to manufacturers, be patient and willing to explain why this flap is here or why that needs to spin—it may not be as obvious to others as it is to you. And don't dismiss their suggestions as soon as you hear them. Think about them—and apply them if they make sense. There is no such thing as bad advice. All advice makes you think and opens your mind to alternatives that may not have been apparent to you before.

ASK QUESTIONS. When you speak to manufacturers, don't be afraid to ask questions. Always ask, "Is there anything else I need to know that we haven't discussed?" Learn as much as you can. If your product will involve a manufacturing process you know nothing about, ask if you can tour the production area so you can get an overview of what's involved. Go to your local library or bookstore and read up on the method.

GETTING AN ESTIMATE

Let's say you're looking for costing information on a sewn product. You look in the *Yellow Pages* under "Sewing-Contract." Sewing contractors often specialize in certain types of sewing, so you need to go through the same process described above for plastics to find out what type of machines, fabric, accessories, cutting, etc., are appropriate for your sewn product.

When you speak to manufacturers, don't be afraid to ask questions.

Once you determine the type of manufacturing needed, it's time to get a *quote*—an estimate of how much it will cost to have your product made. Ideally you should get three estimates, but that may be difficult, depending on where you live; you definitely want at least two.

How to Ask

When you ask for manufacturing quotes, manufacturers will ask what quantity you want. This is a tricky question. If they get the sense that you aren't a "real" potential customer—that you're just asking around for estimates—manufacturers may try to get rid of you with a ridiculous quote.

So when they ask you about quantities, you might say something like, "We would like a preproduction run of 200 or so, and then our orders will range from 2,500 to 10,000." Translation: You want them to manufacture 200 copies of your product before you make any decision on a whole order; then, if those are satisfactory, you will order between 2,500 and 10,000 at a time.

Tailor the Quote to Your Needs

If your product is a new automobile, it would be ridiculous to request a preproduction run of 200. You don't need to see 200 before you decide if a car's being made properly. If your product is a new toothpick, the manufacturer's minimum order will probably be more in the range of 50,000. So tailor your request to your product, using your best judgment. If you're uncomfortable, just be honest; tell the manufacturer you're just at the information-gathering stage. Some will help you out anyway.

If manufacturers get the sense that you aren't a "real" potential customer, they may try to get rid of you with a ridiculous quote.

You want a reasonable quantity cost to present with your product idea to another company (or for yourself). You may also want a quote for the smallest quantity you can get away with for samples, focus groups, etc. Remember, in most cases when you're getting a quote to go along with a presentation for licensing, you won't want to bother with the lower numbers.

Your goal is to get figures based on the approximate retail price of the product. As I already said, if you have a new toothpick, you will probably work in figures between 50,000 and 500,000. If you have a product under $5, you probably want to work in the 5,000 to 50,000 range. If the product is $5 to $15, get a quote in the 2,500 to 20,000 range. Over $15, get a quote for a minimum of 1,000 to 10,000.

Quantity discount. I've given you two quantity amounts for each example because you want to find out how much the price goes down when you order more. This is called *quantity discount.* You could ask a manufacturer for a price for 1,000, then ask where the first "break" is—in other words, how many more you have to buy before you get a better price. It might be 2,500 or 5,000. You may also find that the price for 1,000 is the best price.

Component Purchasing

Component purchasing involves buying all the materials or parts you need and putting your product together yourself—or supplying them to a manufacturer to put the product together for you. Perhaps you have found a sewing contractor who will supply thread, but not elastic or fabric. Or that the contractor you want to work with will supply fabric, but the cost is so high that you'd be better off supplying your own. Ask the contractor to refer you to sources for these other materials—or go back to the *Yellow Pages*.

Your goal is to get figures based on the approximate retail price of the product.

If the manufacturers you call for quotes give you a price that includes everything—all supplies and production costs—you may want to ask for a second quote that covers manufacturing only, using materials supplied by you.

Time Studies

If your product is sewn, the sewing contractor may give you a price based on a *time study*. If so, find out the rate per minute. When I was working on my Little Shirt Anchors, a contractor offered me a rate of 25 cents per minute and estimated that each Shirt Anchor could be made in four and a half minutes. When I sewed one at home, it took me much longer than that. So this sounded like a good deal to me.

To get a time estimate for manufacturing purposes, it's useful to gather materials and make your own samples so you can compare how long it takes you versus how long manufacturers estimate it will take them. Whether this is feasible for you depends on the complexity of your product and the cost of materials. Making 20 to 30 samples of your product is ideal. If you can't make this many, make as many as you can.

Then look at your assembled product and separate the manufacturing into steps. The Shirt Anchors went something like this:

1. Sew tubes of fabric.
2. Thread tube with elastic.
3. Attach garter grips.
4. Attach Velcro.
5. Trim threads.

After you separate your product into steps, start a timer and perform the first step for each of the 20 or so units you're going to make. The moment you are done, check your timer. When you have completed step one for each of your 20 or so units, divide the time by the number of units you worked on, and you have a final time for step one.

Proceed through all the steps just like this, add up the final time for each step, and you will have a time estimate to compare to manufacturers' estimates. Of course, contractors will have a different person work on each step, and each will become extremely efficient at performing that step after doing it over and over—so it's reasonable to expect they will take less time to put a product together than you will.

Other Things to Find Out

PRODUCTION SAMPLES. Ask manufacturers if they can give you *production samples,* and at what cost. For a sewn product, this is usually not a problem. For a product that requires a *mold*, a *dye*, or *tooling*, it may be difficult to get samples. Ask, ask, ask! (Tooling, dyes, and molds are all terms used for an item that needs to be made in order to make another item. The variety of materials, sizes, and uses are too numerous to list, but they could apply to everything from the mold to make a plastic tape dispenser to the dye for punching hang holes in a package.)

SETUP COSTS. Don't be surprised to find out that the costs for molds, tooling, or setup are anywhere from $1,000 to $100,000. If the first manufacturer you call quotes a very high price for these preliminaries, get a few other quotes. But if the price remains high, you may need to consider going back to the drawing board.

GETTING YOUR COSTS DOWN. Each manufacturer that provides you with a quote should hear the same question out of your mouth: *"What can I do to make this product less costly?"* If you aren't asking this, you aren't taking advantage of expert advice. Again, remember that you don't have to take the experts' advice—but you should always listen to what they have to say.

> **Each manufacturer should hear the same question from you: "What can I do to make this product less costly?"**

IMPORTING YOUR PRODUCT

You may find that the costs to make your product in this country are too high for the perceived value of the product. That's what happened with my Little Shirt Anchor. Each one cost about $1.71 to make in the U.S. That would make the retail price somewhere in the neighborhood of $6.84, figuring on a double keystone (meaning, that the store is charging double what it paid you for the product, which was $3.42, or $1.71 × 2). But $6.84 was much too high a price for it.

I decided to look into the possibility of having my product manufactured overseas to save money. You may be saying to yourself, "Not me. My product will be put together by American hands." That's the way I felt at first too. I was adamant about having my product made in this country. But as I learned more about what

(text continued on page 77)

Lesson Learned: Be Ready for the Unexpected

When I first started with Little Shirt Anchors, I sewed them all myself. Now let me assure you, I had no idea how to even thread a sewing machine before I started, but where there's a will there's a way! Soon I was making enough to supply a few dozen at a time to the few stores I was selling to.

Then I decided to attend a Baby Fair, a two-day event open to the public. They stroll through the aisles to see new products and buy what they want. I expected to sell quite a few Shirt Anchors—more than I could sew myself, so I went about looking for a sewing contractor.

I made dozens of calls, took my product to a few sewing contractors for prices, and to make a long story short, finally settled on one. I brought him all my materials: fabric, elastic, Velcro, etc., six weeks before the show. I told him I had made dozens of them myself, so if he had any questions or problems I would be happy to answer them.

It was a great relief to know I didn't have to do all that work myself anymore!

Well, I called him about once a week to check progress. He always told me things were going fine: No problems, no questions, and yes, the order would be done on time.

We got down to the final week, and I was anxious to pick up my Little Shirt Anchors. The sewing contractor was gone Monday and Tuesday. I spoke to him on Wednesday, explaining I was eager to pick up my units because the show started Saturday. He asked if I could come Friday, because he personally wanted to inspect them and wouldn't have time until Thursday. I agreed.

On Friday at 3 p.m. I went to downtown L.A. to pick them up. I had many last-minute things to do before the show that day, and this was my last stop. When I walked in the back door, the first thing I noticed was a pile of bags off in a corner that I thought resembled the bags I had brought him six weeks ago. But that was impossible, I told myself. As I waited for the contractor to come out of his office to see me, I slowly made my way over to the bags.

(continued on next page)

Imagine how I felt when I looked in those bags and found all my material untouched! The first thing I wanted to do was strangle him, but he hid in his office so I couldn't. I picked up my bags and made my way home in rush hour traffic.

I walked in my door in tears at 6 p.m. on Friday and announced to my husband I was going to bed, not to wake me until Monday. He told me not to worry, then quickly called his mother, who came over with her sewing machine and my father-in-law. We all proceeded to work well into the night on a Little Shirt Anchor assembly line. I went to bed at about 1 a.m., exhausted. When I awoke the next morning, my husband was still *at the sewing machine. This man had never sewn in his life before that night; he's just a fast learner.*

The moral of the story? Always expect the unexpected. Always double-check and have backups. Always be kind to your friends and relatives, because there will be times when what you need more than anything else is their understanding and support.

---

was involved, I realized that my product could not be made in the U.S. and sold at a reasonable price.

Should You Go Overseas?

You will find that if you mass-produce your product, you can use American labor and still keep your price down. The same may hold true if your product is not labor-intensive—if manufacturing is not more expensive than your materials. But if it is, you have a candidate for overseas manufacturing. My Little Shirt Anchor, unfortunately, fell neatly into this category.

Finding a Foreign Manufacturer: The Import Agent's Role

The easiest way to find foreign labor is through an *import agent.* An import agent will be your eyes and ears in a foreign country. He or she can act as your manufacturer—guaranteeing quality, getting your goods through customs, making sure there are no quota problems, paying duties, you name it.

To find an import agent, ask for references from other professionals you have worked with—attorneys, financial consultant, etc. You can also look in the *Yellow* Pages under "Importers" and "Customs Brokers," or you can check with the import/export trade publications such as *Export Today*.

IMPORT AGENTS' FEES. Agents' fee structures vary. They commonly require a flat price for each piece in your manufacturing order. That price covers the cost of the product as well as associated import costs. This arrangement may be preferable to one in which you have to pay a per-item cost along with additional fees for communications, delays in clearing customs, and that sort of thing. The reason: You have no control over those extra costs; the import agent does.

AN EXAMPLE: When I started importing my Little Shirt Anchors, one of my shipments was delayed in a customs warehouse in Los Angeles for five and a half weeks. The government charged a storage fee for that time, which my agent had to pay. During all those weeks, my agent sent communications overseas daily, trying to get the appropriate paperwork for customs clearance. Because my fee arrangement was per piece, I had no additional costs.

I realized that my product could not be made in the U.S. and sold at a reasonable price.

Doing It on Your Own

If you'd like to explore importing without an import agent, you first need to find the manufacturers. Several countries have importing offices or associations, such as the Hong Kong Trade Development Council and the Japanese Export Trade Organization. To locate the trade organization for a particular country, contact its Embassy or Consulate General office. Many countries have offices in larger U.S. cities; you can always start by contacting the office in Washington, DC. Or check your local library.

One word of caution: Make sure whomever you contact is really associated with that country's government. Many private companies

---

A Lesson Learned: Grow Slowly and Steadily

I have a very good friend who owns a company called Flap Happy. She is absolutely the best example I can give for not growing too quickly, so that you can see ahead to the consequences of your decisions.

She started with one hat, the Flap Hat, and sold it through specialty stores. The hats sold very well and she began to get a lot of attention from the mass merchants. But in order to supply them, she would have to have her hats imported because of cost. She has great pride in the fact that her products are manufactured in the U.S. But more important than that, she recognized that the extra work involved with importing and selling to the mass merchants, even though the revenues would be tremendous, would mean she could no longer offer the type of personalized customer service that had made her so successful to begin with. She would be a large, faceless company that put cost before quality. She would lose her special touch, and many of her smaller specialty stores.

She has expanded her product line now to many styles of hats and an entire line of children's clothing. She sells worldwide and has a multi-million-dollar business, which she started from one room in her house, with one product. Her firm appeared in Inc*'s 500 list in 1994 as company #194.*

And she did all this by keeping focused and knowing when to say no to expansion. It hasn't hurt her one bit, and I think of her often when I find myself getting carried away with the potential of something instead of concentrating on one area and giving it my best.

---

use the name of a foreign country to make you believe they are in some way affiliated with that company's government.

ANCHORS AWAY

Here's how I went about finding overseas labor: I targeted Taiwan, Japan, and Hong Kong as likely candidates and set about contacting each country's trade consul. I had no luck locating Taiwan's office.

Japan, it turned out, was a good supplier for plastic and electronics but wasn't right for my Little Shirt Anchor. The Hong Kong Trade Development Council told me to send a sample of my product to be displayed in the country; any manufacturer interested in it and able to make my product would contact me.

Several weeks went by, and I didn't hear anything. Then one day I received a padded envelope in the mail, postmarked Hong Kong. Inside were sections of different types of elastic with little tags on them. The tags all had what looked like pricing. I assumed I had gotten on some kind of mailing list due to my earlier inquiry, and this company was sending me samples.

It's a good idea to use both an import agent and a quality-control company, at least until you've got your feet wet.

I dumped out the contents of the envelope on my desk—and my eyes nearly fell out of my head. Among the little samples of elastic material was a crudely made "sample" of my Shirt Anchor! I deduced that the company took the sample Shirt Anchor I sent to the trade development council, figured it must be something of value to people in the U.S., and started including it in the sample packages they sent out. I knew they were knocking me off because the Shirt Anchor was not called out to my attention in any letter to me and a textile and trim distributor whom I had worked with received an identical envelope with the same sample.

I faxed a very nasty letter to the company, to no avail; I received a letter back that indicated the company either didn't understand or was pretending not to understand that I was angry. The letter included quotes for quantities and delivery times and concluded by saying the company was happy to do business with me.

My practical side got the best of my anger. The company's price quote for labor was .36 cents per Little Shirt Anchor; aside from the fact that they'd tried to rip me off, the price was very appealing. I wanted to give the company my work.

The Import Agent Steps In

It was clear to me that if I wanted to proceed with this arrangement, I'd need help from an import agent, who would have experience communicating with overseas businesses.

Through sheer luck, I was referred to an agent with whom I still work with to this day. He contacted the company in Hong Kong and set up an exclusive manufacturing arrangement. I would buy the product from that company exclusively—if they agreed not to sell my product to anyone else. The deal worked out fine, and the company still makes the Shirt Anchors to this day.

An exclusive agreement with a foreign manufacturer isn't really worth the paper it's written on.

I must note here that this type of exclusive agreement with a foreign manufacturer isn't really worth the paper it's written on. What would I do if the company did sell my product to someone else? Fly over there and throw a fit? I don't think so. The reason you make the agreement is that it is a deterrent, just like a patent, trademark, or copyright. It also encourages them to sell only to you, so that you buy only from them. The legal fees would be completely unreasonable, as you would need counsel here and in any foreign country. Added to that are all the communication fees, foreign court costs, etc. The patent office offers some protection services for copyrights, trademarks, and patents if recorded with them. For information write to: Commissioner of Customs, Attn: IPR Branch, Room 2104, U.S. Customs Service, 1301 Constitution Ave., NW, Washington, DC 20229.

I was just very fortunate that the company I work with has chosen to uphold the agreement—thanks in great part to my import agent.

THE DOWNSIDES OF IMPORTING—HOW TO AVOID THEM

You've probably heard horror stories of people who ordered a product from overseas and ended up with a poor-quality or defective item. And they were stuck with it. That's not the only risk involved in importing your product; but a few precautions should help you avoid the bumps in the road.

Quality Control

To prevent quality problems, my import agent employs a *quality-control company* that operates worldwide. Another expense, but it ensured that what I ordered would be exactly what I got, down to the number of stitches per inch, the grade of thread—every detail. It's a good idea to use both an import agent and a quality-control company, at least until you've got your feet wet.

It's mandatory to see production samples if you're having printing done overseas—prints and ink colors are different from those in the U.S.

It's mandatory to see production samples. This is particularly important if you are having printing done overseas. Stock prints (stock prints are printed materials that are generic and therefore can be used for any purpose by anyone) and ink colors are very different from those in the U.S., and print quality is often poor. Unless you are manufacturing large quantities, or using one of the few manufacturers that have high-quality printing equipment, you may be disappointed. If possible, have all your artwork, negatives, and film prepared here, and then send it overseas with color samples to your manufacturer.

Minimum Quantities

One common drawback for new inventors who want to import their product is that the minimum quantities are usually quite high. Don't be surprised if, for example, you get hit with a minimum quantity of 5,000 to 25,000 units on a product that costs, say, $1 per unit to manufacture.

Think carefully before agreeing to this sort of commitment. I would not recommend ordering such a large quantity of goods until you have a healthy number of existing sales. Existing sales are not calculated on promises from people who say they will order your item or tell you confidently, "You're going to sell a million of these things." Existing sales are calculated on, quite simply, *the number of products you have already sold,* using local suppliers. If your sales have been brisk, it may be a safe bet to go ahead with an importing arrangement. But aim for a reasonable minimum quantity—not something in the high thousands.

The company I work with once received a new product sample that everyone really liked—except for the dreadful design on it. I called the product's creator to tell him we liked the idea and were thinking of going ahead with it. Then I began asking him questions about manufacturing. When he told me he had 100,000 units in stock, all with that awful design, we were no longer interested in the product.

PAYMENT

Paying for overseas manufacturing is not simply a matter of sending a check and waiting for delivery. Nor will a foreign company manufacture goods for you, send them over, and await payment from you.

Letters of Credit

A *letter of credit*, or *LC*, comes in handy when paying for overseas labor. Here's how it works: Let's say your order is going to cost $10,000. You take the money and put it into a bank account you cannot withdraw from. (You can earn interest while the money is in the bank, provided, of course, you put it into an interest-bearing account.) Or you have a $10,000 line of credit with the bank, which it holds for collateral.

The bank then issues an LC to your manufacturer, stating that the bank has the money, will hold it for 90 days (or whatever length of time you agree to), and will then release it to the manufacturer upon receipt of certain documents—such as a record of shipment and a statement from your quality-control company that the goods have been inspected and have met your standards. (A quality-control company will do its last inspection just before the goods are put on the ship, right at the dock.)

If everything is in order, you get your merchandise and your manufacturer gets its money.

LCs can be used with domestic as well as overseas companies. In any case, they provide mutual protection and are worth the fee charged by banks. Not all banks handle LCs, but your local bank should be able to refer you to one that does.

CHAPTER SIX

LICENSING YOUR PRODUCT

Skip the Blood, Sweat, and Tears, and Go Directly to the Bank

When you're ready to unleash your product on the world, the easiest route to take is licensing it to a company for a royalty. Under a license agreement, you get a certain sum of money for each unit the company sells.

When companies consider licensing a product, they look at several factors. Before presenting your product to anyone, you should look at it with three criteria in mind.

WHAT COMPANIES LOOK FOR

• **Is it unique and patentable?** This is the first thing a company will look for.

• **Does it fit into the company's existing product line?** Can it sit on a shelf next to the company's other products and look like it fits in? If it can't, you shouldn't be presenting your product to that company in the first place. (See later in this chapter for guidance on locating companies that may find your product a good fit with their line.)

• **Can it be manufactured for a reasonable price?** The company will figure out what it will cost to make the product and what they will have to charge their customers. If you want your idea to get licensed, don't invent a hovercraft-type product. The hovercraft is an efficient way to get around, but you don't see too many of them because they cost $200,000 per unit.

When a company looks to license a product, it wants something its designers and researchers haven't even come close to before.

If you follow my advice in chapter five about doing research on pricing and production, you will know up front if your product can be made for a reasonable price.

Is it One of a Kind?

Uniqueness is the first thing they'll look for. As I've mentioned before, the fact that you've been to every store in your area and have never *seen* anything like it doesn't mean there *is* nothing like it. Believe it or not, a few years ago five different clients brought me the same "new" product idea—kneepads for babies—within about six months. As it turned out, baby kneepads were already on the market.

Very few products that simply do something a little better or more efficiently than a similar product are candidates for licensing. When a company looks to license a product idea, it wants something really different. Something its designers and researchers haven't even come close to before. Something that screams with uniqueness.

A company will also be very reluctant to pay for a product idea that's simply brought in from another market. Baby kneepads are a good example. Taking an existing product, a kneepad, and adapting it to fit babies is not a work of imaginative genius. I don't in any way mean to insult those who came up with this idea. Baby kneepads are a good product, and they've sold very well. I'm speaking strictly in terms of licensing potential here.

Is Your Product Patentable?

The biggest question companies ask themselves is: "Can this product be patented? As I mentioned in chapter four, companies will generally require a patent for any new product they accept, in order to prevent rip-offs. They aren't going to pay you royalties on a product that is not protected.

The biggest question companies ask themselves is: "Can this product be patented?"

I've actually done many product presentations before filing a patent, but I don't tell the companies this. However, I have never presented an idea to a company without first consulting my patent attorney. I know patents are costly. I don't want to spend the money if I don't have some assurance of potential revenue from a product idea. But I do go to the limited expense of having my attorney evaluate the idea first. I may even have a patent search done so I'm certain I'm not wasting my time. (This process is discussed in chapter four.)

Recently I made a product idea presentation to LA Gear. The company loved my product. Of course, I had done all my research, so I could give the reviewers information on costing, market testing, and focus group testing. I even provided them with samples of the product that had been used for several months, to show how durable it was.

The product filled a definite need, it was new and different, it was, in sum, all the things a company loves. There was only one problem: The idea was so simple that LA Gear was afraid of putting a lot of money into manufacturing it and launching a promotional campaign, only to have some other company bring out something similar six months later that would provide them with stiff competition.

The patent on a new product must be very strong or companies simply won't want to take the risk.

Choosing a Likely Company

When you're looking for a company to license your idea, the first one that comes to mind will probably be the biggest in the industry. You have a baby product, so you head right to Fisher-Price or Playskool. A new tool, Black & Decker. Shoes, Reebok. And so on.

Put on your nonrose-colored glasses—the ones that help you look at your product objectively.

We all want big sales so we can get big checks. But many times the biggest company is not your best bet.

Know the Product Line

The first thing you need to consider when evaluating companies is their *product line*—what they sell—to determine whether your product fits in. Put on your nonrose-colored glasses—the ones that help you look at your product objectively. See what is there, not what you wish were there.

Let's say your invention is a new kind of wind-up screwdriver. Now you need to find a company to license it. You figure your product is a tool, so you take it to Black & Decker. However, Black & Decker's strong area is *power* tools, so chances are very good the company will turn you down. And it does. "But it's a great idea," you say. It doesn't matter, your product is outside Black & Decker's market and doesn't fit the company's product line.

Research Company Catalogs

Before submitting your product to a company, call and ask for a copy of its catalog. (Of course, you may already have one from a while back, when you were first researching your product idea.) You want to make sure, first of all, that your product really does fit in with the rest of the company's line. For instance, maybe you've designed waterbed sheets, and you've targeted a home furnishings company that seems likely to carry that sort of product. But when you receive its catalog, you discover it only makes tablecloths. That's an entirely different category of sales; it's very likely the company won't be interested in your sheets.

Looking through catalogs can help you determine how the product line is structured and what sort of items are most important. If you look through a catalog from Rubbermaid, you may find plastic storage containers on the first few pages and kitchen utensils listed in the back. This indicates what items are most popular with Rubbermaid's customers, as well as which ones the company wants to emphasize.

Make it Your Business to Know Their Business

It's useful to find out who owns your target company, what other companies it owns, and whether any acquisitions or mergers are in the works. Trade publications come in handy here, as does the business section of any major newspaper.

If the vice president of sales from one company moves to another company, she may take her ideas and work style with her. Suddenly a company that did not accept outside ideas may change its policy; or a company that only dealt with products manufactured in the U.S. may decide to begin importing. Maybe a computer software company is buying a computer furniture company. These are all things you should make it your business to know.

Try to avoid submitting your product to a company during a merger. I was once in the middle of negotiation with a company when a merger went through; next thing I knew, I was asked to resubmit my product and start from the beginning with the "new" people. I'll never do that again.

A Cautionary Tale

Here's an example of what happens when you don't research companies carefully before submitting your product:

When I first invented my Little Shirt Anchor, I perused the juvenile market carefully. I had children, I watched TV, and I read parents' magazines, so I felt familiar with all the big companies and decided to approach one of them with my product.

After much consideration, I chose the William Carter Company, a manufacturer of children's clothes. I knew the name meant quality and that the company had been around a long time. I figured its distribution must be tremendous, and I knew it did a fair amount of advertising. These factors were important because I realized that my product, because it was so unusual, would require advertising or growth would be extremely slow.

I have definitely learned the importance of doing my homework.

The first time I called Carter's, I was told the company does not review outside ideas. Well, a good inventor never takes no for an answer. But Carter's is in Massachusetts and I live in California, so I couldn't just drive over and demand to be seen. So I called the company and gave a secretary some ridiculous story about having to send flowers to the head of research and development, whose name I had misplaced, for my boss. Could the secretary give me his name so my boss wouldn't know I had lost it? The "one secretary to another" ploy worked; I got the name.

My first few attempts to call this man were not very successful. I decided, finally, to send him a form letter with a sample. Then I happened to speak with a retired businessman who used to work in R&D, who told me he would never dream of sending a product idea through the mail; he said he used to throw out most ideas that came in without even giving them a glance. If he did happen to review an idea, it had better not be after he'd fought with his wife that morning, or it wouldn't stand much of a chance. He told me to make a presentation in person or not at all.

Well, there it was—a road to follow. But how could I get Carter's R&D honcho to see me from all the way across the country? (Especially when he wouldn't even take my calls!) I started by sending him cards and by leaving messages. Cards that said, "I'm thinking of you, are you thinking of me?" or "Miss you—like to see you soon."

After enough persistence I finally got to talk to him. I then flew to Boston, made my presentation, and left feeling like a winner. He liked the product very much but said he needed to run it through a test panel and source the manufacturing (find a potential manufacturer and get an estimate on cost-per-unit). He would get back in touch.

When he called back, I learned of my big mistake. I had already determined, as you may recall from chapter five, that my Little Shirt Anchor had to be imported because it was labor-intensive—the handling costs were extensive in comparison to the material costs. If it were made in the U.S., the retail price would have been too high for my target market.

The problem was that 99.9% of Carter's product line was made in the U.S.—which I would have known if I'd done enough research. Even if Carter's loved my product, there was no way the company could make it for a price that customers would pay.

This episode definitely taught me the importance of doing my homework. And by homework I mean:

- Read, read, read trade publications
- Obtain any literature the company puts out by requesting it (catalogs, "helpful hints," etc.)

• Call the company contact for inventors and ask if they have any guidelines or product categories they specifically *want* to see ideas for, or that they *don't want* to see ideas for

• Read all product packaging from the company

• If the company is publically owned and traded, call and ask for investor information. Make sure you are on the mailing list to receive current press releases

• Ask other people in the industry about companies you're thinking about contacting

HOW TO APPROACH COMPANIES

Many inventors approach companies with their product by writing a form letter or description of the product, then sending it out at random to all the major companies in their field. This is a waste of time.

Instead of setting yourself up for rejection, set yourself up for success.

As I mentioned before, I've learned that research and development departments of most major companies will not look at an idea for a product when you submit it this way. And the ones that do will usually reject it. For all your efforts you'll most likely end up feeling frustrated and defeated. You don't need that; you need to keep feeling positive about your product.

Instead of setting yourself up for rejection, set yourself up for success. Target specific companies whose product lines are good fits for your product; then call them and ask them how to proceed.

I told you about the retired R&D exec who said product presentations should be made in person or not at all. I've since learned that that advice is a bit extreme. Traveling hither and yon is expensive and simply isn't feasible for many people. As luck would have it, the company that's best for your product is on the other side of the country. And many companies refuse to grant personal interviews to inventors they have never worked with before. So just contact the company and find out if it will review outside ideas and what approach it prefers.

PREPARING A PROPOSAL

Once you've targeted a specific company, you're ready to convince it that your product belongs in its line. In order to do this you need to write a *proposal*, using what you've learned about the company, that makes acquiring the rights to your product the easiest, most painless procedure in the world. You will use this proposal as part of your personal presentation.

Your proposal can't be a form letter. It must be custom tailored to address points of concern to a particular company—assuring it, for example, that the manufacture and distribution of your product will not interfere with policies the company already has in place.

Companies will be impressed that you've taken the time to get to know them.

ANTICIPATE THEIR QUESTIONS—AND ANSWER THEM. When you submit a proposal, you are relying on someone else's ability to judge your product. The more questions you can answer on initial presentation, the more likely you are to succeed. For lack of information on costing and manufacturing, a person reviewing your proposal may decide, "I really like this idea, but I bet it will cost too much to make, and our manufacturing division has a three-month backlog on pricing. Send the inventor a rejection letter."

LET THE COMPANY KNOW YOU ARE FAMILIAR WITH ITS PRODUCT LINE. Using information you've gathered from the company's catalog, say right up front that you have a new product for its "Care and Feeding" area or its "Family Games" area. Use the same names the company uses in its literature. I promise you it will make a huge difference. The person you're submitting to will be impressed that you've taken the time to get to know the company, that you understand its unique marketing techniques, and that you've customized your submission to fit the company's needs.

Information They Can Use

Write a proposal that gives companies information they can use. Basically, they'll consider four things when evaluating your product idea:

- **Marketing:** Is there anything else on the market like it? What is the ideal price point? Is there a need? Will it fit in with our current product line?
- **Manufacturing:** How much will it cost to make per piece? Will there be tooling or dyes required?
- **Legal:** Is this a patentable idea? If a patent already exists, how strong is it? Does it infringe on other patents?
- **Research and development:** Does this product resemble anything we have ever seen, worked on, or discussed, or anything we're currently working on?

Throughout this book you will find details on gathering this submission information. Include as many facts as are relevant, but be brief. Your tone should be professional and unemotional.

In your *cover letter* you need to make it clear you are submitting an idea seeking a license agreement. State the name of the product idea ("garden hose with three-way head" rather than "Handy Dandy Hose Converter") and a brief description. Also state in the cover letter that a *signed submission form* or *nonconfidentiality agreement* is included, if that is one of their requirements (see later for more on this).

A Sample Submission

My product proposals include six components (product description and purpose, manufacturing considerations, market testing, marketing considerations, and suggested name, sizing and product line considerations [as applicable]) and a summary. Take a look at the actual submission that I made for one of my inventions, in the Library of Resources.

Some basics:

- Always identify yourself, including your name, address, and phone number, in the upper left corner of the top sheet of your proposal.
- Identify the person you are submitting to by name, title, company, and date in the upper right corner.
- Do not staple the pages of your proposal.
- On the bottom right corner of each page, write the page number, the total number of pages, and your initials; e.g., "page 1 of 6, JR." (Page 1 is your cover letter.)
- At the bottom of your cover letter, indicate what is included besides documents.
- Make a copy for your files and a copy for your attorney. If you enclosed samples or drawings, make copies or take pictures of what you send.
- At the end of your cover letter, under your signature, type the name of your attorney, i.e., cc: Joe Brown, Attorney at Law, to alert the company that an attorney is involved and will receive a copy of the proposal—but not in a threatening way.

Keep in mind that the people who review outside ideas for a company are not going to share your vision of your product.

PRESENTING YOUR IDEA

Preparation is key to successfully presenting your product to a company. I would never make a product presentation that didn't include cost estimates and the results of market/focus group testing (see chapter two). Consider every angle.

Keep in mind that the people who review outside ideas for a company are not going to share your vision of your product. They won't be able to conjure up a precise mental image of the idea you've presented on paper; they won't be able to forgive a flawed prototype as easily as you. They have not been living, eating, and breathing this product like you have.

Show them your prototype. By now you should already have a workable prototype, which you developed while doing your preliminary research and market testing. (See chapter two for details on prototype development and testing.) The first purpose of a prototype is to test whether your product functions. The second is to serve as a sample for market testing and manufacturing considerations. The third—which comes into play now—is to give the people reviewing your product something they can hold in their hand and provide them with essential information: how it feels, what size it is, how it works, and other specifics.

Speak their language. Learn to speak to product submission reviewers in their own language. Among the first questions in their minds will be, "What's the cost? What is our profit margin?" Remember, they make their decisions based on many things, not just on how much they like a product. So do yourself a favor and get some preliminary work done before you present the idea.

Making a Presentation in Person

If you are fortunate enough to make a presentation in person, you will want to make sure you dress in a businesslike manner. Come prepared with a couple of extra copies of your presentation in case more than one person sits in on your meeting. Bring samples with you and be prepared to leave one or more if asked. Bring documents to back up your claims.

You should not reveal, either in the proposal or verbally, when or if a patent application has been filed, nor bring any copies of any applications or issued patents. Your attorney can submit a letter regarding patentability after negotiations have started. Do let it be known that you have a patent attorney and that your patent attorney has worked on the idea, but don't say that an application has been filed if it hasn't. Again, you'll want to get advice from an attorney on all of these points.

Bring a legal pad for taking notes. Also, if you have business cards, bring them. Be sure to ask for a business card from every single person who sees your idea. If you are introduced and forget a name, ask a secretary or receptionist.

A Presentation by Mail

If I'm submitting a product proposal by mail, I always like to find out who will be reviewing it. I want a name. Sometimes it's difficult to get one, but it's worth making the effort. The best way to start is by calling the company and asking for the director of research and development. You probably won't get that person, but you may well find a friendly person along the way—maybe a secretary or an assistant—who can keep you updated on the progress of your submission.

Dos and Don'ts

- **Don't be a pest!** As someone who now reviews outside ideas, I can tell you that nothing is more aggravating than an inventor who pesters me. Give the company time to respond.
- **Don't argue with anybody!** If someone tells you your product doesn't fit the company's line, or it's too expensive to make, or it doesn't have enough customer appeal . . . don't get belligerent. If you can offer facts, do so; then accept their decision.
- **Listen to what people say.** Those reviewing your product are giving you valuable input. Listen carefully to the reason your product was rejected, then do something about it! Consider it constructive criticism.
- **Don't wax poetic about your idea.** Don't tell companies "Everybody loves it" or "Everybody wants one." They couldn't care less what your family and friends say about your product. They need facts: evidence that your product idea has merit. Tell them that 75 percent of participants in a focus group gave your product a grade of eight or higher on a scale of one to ten (but only if this is true!).

(text continued on page 100)

A Lesson Learned: Don't Give Up Easily!

This is one of my all-time favorite stories. And, hard as it may be to believe it, it is 100 percent true. I was trying to license my Snap Laces and thought about approaching Nike. I contacted the company, and was able to make an appointment with the Kids Marketing Director the following month. I live in Southern California and Nike is in Beaverton, Oregon, just outside Portland.

I flew up to Portland on the morning of my appointment, which happened to be a Friday before a three-day weekend. It was raining. I had planned to stay in a hotel in Portland, but I thought I would drive the 30 miles or so to Beaverton before checking in, just to see where it was and how long it would take me to get there.

Well, I found Nike, all 78 acres of its beautiful property. It was still raining, so I decided to get a room there instead of driving back to Portland. By now it was about 11:30 a.m. and my appointment was at 2:00 p.m. So I checked into this little fleabag motel, and settled in to wait for appointment time.

I had two bags of samples with me, plus my proposal and a notebook. I was confident and totally prepared. I knew I had 45 minutes of Nike's time and I was determined to use every one of them constructively.

I left the motel at 1:45. It took me about seven minutes to get there. I drove past the security post and parked. I did my last little pep talk to myself, gathered my samples, checked my hair and makeup, and proceeded to walk toward the main building.

I'm still not sure how it happened, but about 15 strides away from my car I fell. Now I don't want you to think I'm some kind of klutz or anything, but there I was, walking in and next thing I know I'm on the ground on all fours. My right ankle screamed out in pain as I tried to stand up. Balancing myself on my left foot, I noticed my left knee had a horrible-looking wound that proceeded to bleed down the front of my shin. My pantyhose that used to cover that knee now looked like it had been mauled by a bear; there was a giant hole at my knee and there were runs from my foot up my thigh.

(continued on next page)

Then I looked at my watch. It was five minutes before two. So, what do I do? Try to call to postpone the meeting? Where do I call from and what do I say? I knew the marketing director was busy the rest of the afternoon, and I knew he wouldn't be in again until Tuesday because Monday was a holiday. Cancel the appointment? No way. It cost me $800 to fly up there.

So I hobbled back to my rental car with my right foot/ankle in extreme pain and tried to wipe the blood from my leg with tissues. Then I found some of my kids' Garfield Band-aids in my purse and put three of them over my wounded knee. Then I held my sample bags in front of my leg and limped my way into the waiting area.

Now, the last thing I wanted was for this man to think I was some kind of idiot, or to concentrate more on my injuries than on my idea. He came down to greet me and I smiled warmly, making sure my injured limb and the resulting blood loss were hidden behind my bag. He immediately offered to take me on a tour of the grounds. I graciously declined, explaining I had taken a little fall in the parking lot. His eyebrows went up and his eyes became wide as he started to scan my body for obvious signs of injury—and soon noted my obvious signs of injury.

Did I want medical attention? No. Did I want to file an accident report? Later. What could he do? Forget it, find me the nearest table so I didn't have to walk on my throbbing ankle and let's look at my product. He agreed.

We went to the cafeteria, and he became immediately engrossed in the product. In fact, he missed his next appointment and spent two hours with me.

I was thrilled at this chain of events, but somewhat concerned as I could feel my ankle swelling past the point of recognition. I took my shoe off as soon as I sat down, which in retrospect was probably not a wise thing to do.

It came time to leave, so I thanked him profusely for his time. We made arrangements to follow up and he kept samples to show to others at the company. I assured him I could find my way out, but he insisted on escorting me. I couldn't get my shoe back on, so I bravely limped out of the vast expanse of Nike buildings with two sample bags in tow, one high-heeled shoe on my left foot and no shoe on my grossly swollen right foot. I could barely put any weight on that ankle by this point. What a sight I must have been—all dressed up in a business suit.

(continued on next page)

He was very polite and did not make obvious notice of my condition as he shook my hand and said goodbye. I made my way slowly to my car, agonizing in pain. I drove myself to the hospital using my left foot and was treated for a broken foot and a severely sprained ankle.

I won't go tell you details of the rest of my trip, except to say that I spent a painful night in a fleabag motel, with no food and little sleep. I got back on the plane the next day, with the help of a kind Hertz employee who was also an EMT.

Did I do the deal with Nike? No, I am sorry to say. The terms were not to my liking and we couldn't agree on a compromise.

The lessons from the story? Never, ever stay in a fleabag motel with only one employee and no ice on your floor. Carry plain Band-aids as well as the Garfield kind. And don't give up easily when something big is at stake!

Show them samples that have been used to demonstrate wear. Give them manufacturing costs or color preferences—something tangible. Anything else is just a waste of their valuable time.

- **Be brief, punctual, and professional.** If a company asks for samples by next Thursday, make it happen. If you're asked how you came up with your product idea, answer in 30 words or less. If you are scheduled to meet with or call someone at a specified time, don't be late.

A Timely Follow-Up

When doing your follow-up from a personal presentation, you will want to send a thank-you letter to the person who reviewed your product. This letter serves as documentation as well as a reminder that you're out there, looking forward to hearing from your reviewer.

In the letter, list the date of the meeting and whether the idea was well received. ("I was very pleased to see the positive response to my product idea, the Sulfurless Match, during our meeting on April 1,

1994.") In addition to the reviewer, thank anyone else who saw your idea (of course you got their cards or at least their names, as I recommended) for taking the time to participate in the evaluation. If you left any samples, now is the time to mention that as well.

End the letter with an indication of exactly when you will be in touch again to check on how the evaluation is progressing; this should be about two weeks from the date of the meeting, unless the reviewer indicated otherwise at the meeting.

If you submitted your proposal by mail, make a follow-up phone call about two and a half weeks after submission to see how things are progressing. Make sure, first of all, that the company received your proposal, then ask if there are any questions you can answer or samples you can provide, etc. After this confirmation call, send a letter documenting what was discussed.

Send copies of all correspondence to your attorney.

Submission Forms

Most companies have a strict submission process you need to adhere to. This process varies greatly from one company to another. You will need to contact the company and ask what its submission policies are. Most companies will require you to sign some kind of *submission form* or *nonconfidentiality agreement.* As I explained in chapter four, this agreement is the reverse of a confidentiality agreement: It means that you understand the company will *not* hold your idea in confidence. You are allowing the company to evaluate your idea without making any promises to you of any kind.

The following statements are commonly found in submission forms:

The acceptance of a submission for consideration by the Company imposes no obligation upon the Company. No oral representations or agreements by anyone shall be binding upon the Company.

You understand that in the absence of a written agreement executed between you and the company, you must rely solely on the patent, copyright, and other laws affording protection to intellectual property rights for the protection of your submission.

It may be necessary to have the submission examined and investigated by various employees and outside consultants to determine the novelty, unobtrusiveness, practicability, and usefulness thereof, and therefore we will not be bound to treat any submissions as secret or confidential.

The company may also ask you to sign a *nondisclosure statement*, which generally says that while the company will not intentionally steal your idea, you realize it's possible that the company's research and development department may already be developing a similar product. By signing you are acknowledging that this could, indeed, happen, without malice on the part of the company. It is reasonable for the company to ask you to sign this statement before they look at your idea. But have your attorney read it first.

Submission agreements were not created to enable companies to rip you off.

Signing Your Rights Away

Are you wondering why you have to "sign your rights away" in order to make a submission? In general, it's not because companies are out to rip you off. They are simply trying to protect themselves. Many companies have had negative experiences with inventors who either falsely claim to have made a submission or submit product ideas that are very similar to something a company is already working on. If a company is going to accept outside submissions, it needs to protect itself. But again, get advice from an attorney before signing anything.

Don't let signing a submission agreement stand in the way of submitting your idea. Once you've submitted, the company may still say no, they may try to steal your idea, *or* they may make you rich. But one thing's for sure: If you don't submit to anybody, you'll never have a chance to make a penny.

My advice? Protect yourself before going in by meeting with an attorney, then go for it. I have done literally dozens of product idea submissions, and I haven't had any trouble yet. Of course, my presentation has something to do with that. If you act like you know what you're doing, you're on much steadier ground.

How to Take "No" for an Answer

If you end up getting a rejection from the company, which by the odds is pretty likely, use it to your advantage. Really listen to the reason they give you for rejecting your idea. If the cost was too high, then your next step is to re-evaluate your product idea in order to bring the cost down. You might consider redesigning it to use different materials or less materials.

It's extremely rare for a product to make it to market without going through a modification process.

If you're told it is too similar to other products on the market, find out which products these are. If the company won't name them, do some research on your own. If you haven't evaluated your market thoroughly, *do it now!*

If the company says your product infringes on an existing patent, contact your patent attorney *ASAP* and get a search done to see what they are talking about. Evaluate the patent and redesign your idea if necessary. Again, don't be so in love with your product that you are unwilling to change it. You may need to in order to make it succeed. It's okay if the final product is different from the first design. In fact, it's extremely rare for a product to make it to market without going through a modification process. The original prototype is just a sample, something to base further design changes on.

What If You Get a Rejection and Aren't Given a Reason?

That does happen with larger companies that receive several thousand submissions per year. They simply don't have time to reply individually. If it happens to you, don't fret; it's nothing personal. It's probably not worthwhile to call the company and ask why your submission was rejected. You'll probably just irritate someone, who may well come up with any old answer just to get you off the phone.

However, if you do call, don't be argumentative. Be polite; thank whomever you talk to for taking the time to answer your questions; and indicate that any insights he or she can provide will be very helpful. If you are careful not to put anyone on the defensive, you'll get much more information.

What if you *didn't* get a rejection from the company, and they want to license your product?

Congratulations!

They will send you an agreement for review and signing. This brings us to our next discussion.

A license agreement allows one party to use something that is patented or trademarked by another party.

WHAT IS A LICENSE AGREEMENT?

A *license agreement* is a contract between two parties, the licenser and the licensee, allowing one party to use, for a fee, something that is patented or trademarked by another party. There are many different types of license agreements—for characters, for names, for people, and for many other areas of commerce. This chapter will focus on product idea licenses.

A license agreement is usually struck between a company and an individual for use of a patented idea. An example: Let's say you are the licenser and Ford Motor Company is the licensee. Your license agreement might state that you are giving Ford permission to use your idea on their cars in exchange for a fee or royalty that is paid to you each time the company sells a car that incorporates your invention.

The details in a license agreement are usually quite lengthy, covering areas such as how and when payment is to be made, exactly how the company is allowed to use your idea, which party will pay fees associated with patent infringement (someone else using your idea), minimum commitment on the part of the company, and more.

A Sample License Agreement

To give you an idea of how such agreements are worded, I've included in the Library of Resources an excerpt from my first draft of agreement to license my Little Shirt Anchor to A-Plus Products.

One license agreement is entirely different from another, so you will have to consult an attorney to write one that suits your circumstances. In my case, you notice several mentions of the fact that I had existing accounts and inventory. My lawyer and I viewed this as an asset, so we added it to our agreement. We also felt that my product knowledge was an asset, so it was included.

It all boils down to this: You can ask for anything you want in a license agreement. Whether you get it is another story.

You can ask for anything you want in a license agreement. Whether you get it is another story.

NEGOTIATING A DEAL

When it comes to negotiating the financial terms of your agreement, there's one near certainty: You won't get exactly what you start off asking for. The process is similar to buying a car. You settle on a price based on the type of agreement and options you want. You may have to give up one point to gain another. For example, in my letter of intent for the Little Shirt Anchor, I asked for quarterly guarantees against royalties. The company didn't agree to this request as initially worded, but I did end up with a smaller guarantee.

Payment Formulas

Many companies have a formula for determining the financial terms of license agreements. Often this formula includes royalty payments based on *net sales*. The term net sales can mean many things. It may mean all sales, minus returned items. It may also reflect deductions for manufacturing costs, sales reps' commissions, and "administrative" costs, as well as freight and advertising expenses.

For example, say you get a 2 percent royalty on the net sales of your pencils. Each pencil is sold for $1, and the company sells 1,000 pencils. Let's figure what you might get if we assume your royalty payment is based on net sales, and those net sales are determined to be all sales minus returns.

You start with $1,000 in sales—1,000 pencils for $1 each. If you subtract 25 pencils returned from customers, that would now make your net sales $1,000 minus $25, or $975. Your 2 percent royalty would be 2 percent of $975, which is $19.50.

Now imagine you get a 5 percent royalty on the net sales of your pencils. Start with the same 1,000 sold at $1 each. But this time net sales are determined to be all sales minus costs and commissions; that's manufacturing costs of 50 cents per pencil, sales reps' commission of 10 cents per pencil, and freight costs of 7 cents per pencil.

Every detail of your agreement is critical. You need to know what you're doing every step of the way.

Add together these three additional costs for a total of 67 cents. Subtract 67 cents from your starting price of $1 per pencil and you're left with 33 cents. Multiply 33 cents by 1,000 pencils and you have $330 in net sales. A 5 percent royalty based on this definition of net sales would only pay you $16.50.

Surprised? A 5 percent royalty sure *sounded* like more than 2 percent, didn't it? You can see how critical every detail of your agreement is. You need to have a clear understanding of terms every step of the way.

Up-Front Money

You can ask the company licensing your product to reimburse certain expenses—such as attorneys' fees directly related to patents and trademarks—in the form of up-front money. You may also be able to retrieve some research and development costs, but don't expect too much. The company will incur its own costs in this area and won't be eager to pay yours as well. If you tell the company you

Example 1.

2% royalty on net sales

(net sales = all sales − returns)

1,000 pencils × $1 each =	$1,000
− 25 pencils × $1 each =	− 25
net sales	$975

$975 × 2% royalty = $19.50

Example 2.

5% royalty on net sales

(net sales = all sales − costs and commissions)

1,000 pencils × $1 each =	$1,000
−1,000 pencils × .67 each =	670
net sales	$330

$330 × 5% royalty = $16.50

spent five years and $100,000 finalizing your great new shovel, you may very well be told the company has no intention of paying for your mistakes.

If you provide the company with prototypes, you can request reimbursement for these costs. The same is true for artwork or illustrations. You may also be paid for providing consulting services (as an expert on the product) to the company, and you can request part of this fee up front.

Any monies received in advance are usually paid *against royalties.* Thus, if you get $5,000 up front, $5,000 will be deducted from your first royalty check. This advance money is cash in the inventor's pocket right away, and, more importantly, it gives the company an incentive to do a good job promoting and marketing your product so that it can recoup its $5,000 investment quickly.

If your product is expected to reap big revenues, you can ask for an advance against royalties upon signing the agreement and on each anniversary of the agreement.

CHAPTER SEVEN

SUPPLYING THE MASS MARKET

Packaging and Selling Your Product

The alternative to licensing your product to a company is selling it yourself. Now, you may think that after all the work that's come so far—finalizing your product design, determining your manufacturing method, identifying your target markets—getting your product to your customers will be a piece of cake. Once you've got a garage full of puncture-proof tires, the hard part is done, right? Nope. The real work is just beginning.

The ABCs of Sales Structure and Distribution

A manufacturer is someone—like you—who makes a product to sell to others. You will buy materials at a wholesale price from materials manufacturers and put them together to make your product.

If you don't plan to sell directly to the consumer (through a mail-order ad, for example), then you will sell your product to stores or catalogs at your *wholesale price*. The difference between what it costs you to make your product and your wholesale price is your *gross profit*.

The other type of sales are *retail*—the price the end user, or consumer, pays for the product, and is when the sales tax is added to the sale (in states where there is sales tax).

Choosing Your Product Packaging

The package your product is displayed in can be as important as the product itself. It is the only selling tool you have once the product is sitting on the shelf in a store, next to all the brand-name products your potential customers will already be familiar with. You have just a fraction of a second to make a strong enough impact with your product packaging to encourage those customers to examine your product more closely.

The package your product is displayed in can be as important as the product itself.

Conversely, if your product is to be sold through the mail, either through a mail-order catalog or on your own, your packaging should be designed to keep the product clean and protected while being shipped. There is no need to try to "sell" the customer with the packaging because the sale has already been completed. As you read through this section, keep in mind that different sources of distribution have different packaging needs.

Packaging Size and Color

Packaging size should be one of your first considerations. If your package is much larger than the product, stores will refuse to sell it; excessive packaging takes up room that could be used to display other products.

The color of your packaging will also play a key role in how well your product sells. There's no consensus on which colors are most effective, but the goal is to make your product stand out in the crowd—without being offensive. Chartreuse and mustard yellow are probably not the best choices. You want colors that are appealing to the eye.

As a rule of thumb, red is the most eye-catching color to use in packaging or advertising. Look at the logos or packages the next time you are in a store. Red is often used when you want to get someone's attention. Also, if you have a room full of small children, and place a pile of multicolored crayons on a table and tell them to color, the red crayons will be gone first.

When my partner and I designed our packaging for our pet products company we went with red and black on a white background. We used some illustrations of animals that were custom made for our packages. We've had many compliments on the packaging. And when the company I work for now needed a new logo, we went with red.

Pay attention to how your competition is packaging its products; this will help guide you.

The color of your packaging will play a key role in how well your product sells.

Header Cards

Header cards are a very popular and inexpensive packaging option. A header card is a stiff card that is printed with all your product information, your company name, and sometimes instructions, if space permits. A header card can be used in several ways. You can, for example, fold it in half and attach it to a plastic bag containing the product; place it inside a bag with the product; or use it alone, attaching the product directly to the card.

Blister Packs

Blister packaging also involves a card, similar to a header card, made with heavy paper or heavy stock. A plastic bubble or blister is attached to the card or folded within it, with the product inside. You may have seen this type of packaging used for batteries, nails, or baby pacifiers.

Blister packaging can be costly because it requires a machine to attach the plastic to the card using heat, as well as a special coating on the card itself. In addition you must purchase not only cards but blisters as well. Often these are custom made to fit the dimensions of the product, which adds to the price.

Boxes

Boxes are more likely than other types of packaging to be damaged and returned. The secret to preventing this is to make your boxes with a very heavy stock, which costs more—the thicker the paper, the more expensive it is. Boxes are good for packaging gift items, since they can be wrapped easily, and they have an appeal that blister and header card packaging lack.

Tags and Labels

If your product does not require packaging, you may want to use a label or a hang tag. A label is useful for larger items such as furniture or equipment—awkward-shaped pieces that make packaging difficult. You can use either a self-adhesive label or a hang tag (that pesky little tag that's often attached to clothing with a thin piece of plastic).

Durability

You've got to package your product so that it is not easily damaged. As a consultant I had a client who produced small ceramic sculptures with a light inside. They were absolutely beautiful, and my client spent many weeks adjusting the thickness of the ceramic so the light shone through just right. But the packaging was deficient, and hundreds of damaged pieces were returned to the manufacturer. It put my client out of business immediately.

Keep in mind that your product must withstand lots of handling—from warehouse to shipping, to storerooms, to store shelves, to the customer's home.

Appearance

Once again the most important thing to remember in packaging is that in many cases it is the only tool you have to sell your product to your customer in the store. Don't skimp. The customer needs to know several things in a split second: What the product is. How it works. Where to use it. What it's called. You want to make these points zoom out in front of their eyes. They won't take the time to hunt for them.

Look in your local stores for packaging ideas. Look at similar products in the same market. What appeals to you? What looks cheap? What looks professional? What makes you want to buy and what does not? Then decide.

Printing Basics

Four colors are used in the process of color printing: magenta (red), cyan (blue), yellow, and black. Virtually all other colors can be created from these four. Paper is fed into a printing press that is loaded with magenta ink and that color is printed. Then the machine is cleaned and the paper is run through again with the cyan ink, and so forth with the other two colors.

A color separation can be very expensive—but very effective.

How does the printing press know which colors go where? If you want to print a color photograph, you need to have a *color separation* made. In this process a machine creates four different images of the picture; each image is only one color. If you just look at the yellow image, you'll see only the portions of the picture with yellow in them. If you place the blue image on top of the yellow, then you'll see all the yellow, blue, and green (blue and yellow make green) portions of the picture. When you add the last two colors, you will see all the colors in your photograph.

The cost of a color separation depends on the size of the picture. A separation for a three-by-three-inch picture might cost you around $90. Then you have to pay for the actual four-color printing on top of that. So it can be very expensive—but very effective.

If you want a black-ink drawing and words printed in black on your package, you'll get *one-color printing*. If you want a drawing in black and words printed in blue, you'll get *two-color printing*, and so on (up to four colors, which, as I've explained already, will give you the entire color spectrum). The more colors you use, the higher the price.

Pricing Out Your Package

Your local *Yellow Pages* is a good source for all the help you might need in designing your package, from printers to paper companies, illustrators to typesetters. You might also check with a local printer or quick-print shop for referrals of this type. Be sure to shop around for the best price and ask to see previous work.

Before you commit to a package designer or printer, get an estimate. Agree on a price—in writing—and agree on the conditions under which that price might change. After you have the estimate, determine an *acceptable* ceiling. This is the most you will permit the designer or printer to go above the original estimate. You don't want to have an estimate for $135, then get a bill for $500 because the estimate was a "little" off. And be sure you know which changes you will be charged for making after the project begins—and how much they'll cost.

PROMOTIONAL LITERATURE

You'll want to create some type of literature for those who are interested in learning about your product, like store buyers, and consumers at shows or fairs. Should you later decide to license your product, you will want to send such literature to companies that express interest in licensing it.

Brochures or Catalog Sheets

A brochure or *catalog sheet* (don't be misled by the name—it often contains only one product) answers questions about your product's

UPC Barcodes

When designing your packaging, you would be wise to include a Uniform Product Code barcode, or UPC. This barcode is read by an electronic scanner and is used by retailers as a means of keeping track of inventory and generating a customer receipt at the time of purchase.

While not all of your accounts require UPC codes on their product packaging, it is easier to put the UPC barcode in place at the time the package is printed. This will save you from having to ad stickers to the product at a later date if you sell to an account that requires one.

In order to use the UPC barcode system, you need to contact the Uniform Code Council at 8163 Old Yankee Road, Suite "J," Dayton, OH 45458, or call them at 513-435-3870. They will send you an application, which you return to them after filling it out and enclosing a fee. This will get you a manufacturer's number, which will always be the first set of numbers appearing in the UPC barcode.

The council will also supply you with all the additional information you will need to utilize the UPC barcode system for your product(s).

appearance, size, usefulness, and colors or designs. Unless you want to send out a lot of samples, these are a worthwhile investment.

Your brochure or catalog sheet should include either a drawing or photo of your product. Since four-color processing is so expensive, you may prefer to use an illustration, which is fine as long as it has a professional look to it and isn't amateurish.

Price Sheets

It's probably best to keep pricing out of your catalog sheet. Instead, make a separate price list that can be easily and inexpensively changed. If you create the *price sheet* yourself on your computer, you can avoid paying the printer every time you need to make an adjustment.

Your price sheet should include:

- standard price per unit
- the number of units per box
- the size and weight of the box
- quantity discounts
- your terms (see below)
- where the product will be shipped from, known as the *F.O.B.* (freight on board) point; this is necessary information since purchasers will be paying the cost of shipping your product to them
- UPC barcode (See box)

Terms

Your *terms*, meaning payment terms, will more than likely be net 30, which means the invoice must be paid within 30 days. Sometimes you'll see terms that say net 30/2% 15. This means purchasers have 30 days to pay the invoice, but if they pay it within 15 days, they get a 2 percent discount. You'll want to include a time limit on making claims, such as 15 or 30 days; you must be notified within that period if the product is going to be returned to you because of breakage, damage, or a shortage for example. Chapter ten includes many more details on invoicing and paperwork.

REACHING BUYERS THROUGH DISTRIBUTORS

Buyers work for the stores or catalogs and decide what merchandise the company will sell. These are very important people to know. It's up to them to determine whether your product gets placed. In small stores, the owner often does the buying. Chain stores may have dozens of buyers, each one responsible for one particular section of merchandise for the store.

But there's often a middle player between you and the company you're pitching your product to: *the distributor.* For an inventor with one product, working through a distributor is often the best route to take, unless you plan to add products to expand your line. A distributor can handle many of the problems associated with being a one-product company.

The Distributor's Role

Here's how distributors fit into the picture: You sell your product to a distributor, then the distributor sells it, or distributes it, to places where the consumer can buy it, such as specialty stores, mass merchants, department stores, catalogs, and mail-order houses.

Distributors hire sales reps to represent their line of goods. The distributors handle all the invoicing (billing) and collections for the accounts they sell to; they absorb the losses on accounts that don't pay. In many cases they will attend trade shows with your product and work with companies to offer your product as a premium or in a sale with another product. Some distributors will even do advertising for your product.

A distributor can handle many of the problems associated with being a one-product company.

How Distributors Benefit Inventors

One of the most important things distributors offer is an opportunity to make your product part of a *product line*. This means it's marketed along with several other products, instead of being marketed all by itself—one product from a little company no one has ever heard of.

As I've noted before, being a single-product manufacturer has real drawbacks. It costs the same amount of time and money for a major account, such as Walmart, to set you and your one product up as an approved vendor as it does to set up a company with 100 products. And of course a store has a better chance of reaping big profits from 100 products than it does from your one product. In fact, many major chain store buyers have been known to refer single-product manufacturers to a distributor if they have any interest in the product.

Being a single-product manufacturer also highlights the fact that you are the new kid on the block. Nobody hangs around long with one product, and the trade knows this. As a single-product manufacturer, you're advertising your inexperience and lack of knowledge, and that means potential problems for stores when it comes time for you to supply the product.

Let's say your product is a big hit. Chances are you have neither the capital nor the experience to keep the stores full of inventory. That makes the store look bad and it loses money. And you haven't experienced anger until you have heard what your catalog account says to you when it runs out of inventory and you can't supply any more units until next month.

Many larger accounts choose not to deal with a small company that has only one or two products. It's as simple as that.

Good distribution of a product is an essential part of its success. Whether you do it yourself or through someone else, it's going to be a costly undertaking and will require planning and research.

Being a single-product manufacturer highlights the fact that you are the new kid on the block.

Finding a Distributor

If you decide you want a distributor to sell for you, start by checking with trade organizations in your product field. They will usually be able to provide you with a listing of distributors who carry products like yours, just as they can provide you with the names of sales reps, trade shows, and other information related to your industry.

When you contact distributors, try to determine fairly quickly how much interest they really have in your product. Ask them what other products they distribute; this will help you determine if your product "fits." You'll probably want to ask them questions about their customer base—who they sell to—and the number of accounts they have, but don't be surprised if they are reluctant to answer either one of these questions in any detail. To some extent, this is considered proprietary information.

Pricing Your Product for Distributors

One of the first things a distributor will want to find out from you is how your product is priced. In chapter five I talk about pricing structures. You can use those guidelines to help you determine the price of your product, but if you get a distributor involved you may have to adjust that price.

You will need to discount your product enough so that the difference between the distributor's purchase price of your product and its selling price is enough to cover its distribution costs and have some money left over for itself.

Don't price yourself out of the market. Don't take off 10 percent of your retail price for the distributor and assume that's enough. You need to leave the distributor enough gross profit to offer discounts, pay sales reps' commissions, pay for advertising, literature, and trade shows, cover administrative costs such as telephones, office supplies, and accounting, handle bad debts that cannot be collected, and make a profit!

Don't price yourself out of the market.

What *you* have to do is get the product made, ship it to your distributor, and collect a check. Of course, you need to make a profit also, but nobody is going to make any money from nonexistent sales of a product that is too expensive. So be realistic when setting your price.

CALCULATING PRICE. Here are some general guidelines. Keep in mind as you read this that specifics will vary, according to your product, your market, etc.

Assume that your product costs $10 to manufacture and package. You have no distributor; rather, you're selling directly to retail outlets, so, using the pricing guidelines in chapter five, your cost to stores would be about 2 x $10, or $20 per unit. Simple enough.

But if you should decide that you want to work through a distributor, before you price out your product you will need to determine the margin your distributor will want to put on it. If, for example, your distributor wants a 30 percent margin, it will have to buy your product at around $14.

If you know what margin a distributor will ask for on a product, and you want to determine at what price you must sell it to the distributor, follow these steps: First, figure the total cost of making, packaging, and shipping the product, then figure how much profit you want to make on each unit you sell.

Say you've got an item that costs $7 to make, and you want to walk away with $3 on each one you sell. Add these two figures together: $7 + $3 = $10.

This would be the distributor's price.

Now multiply that amount by 1.43 to determine a 30 percent margin on the product:

1.43 x $10 = $14.30.

This would then be the wholesale price, or the product price your distributor would offer to the stores.

Keep in mind that distribution of your product is an essential part of its success.

(1.43 is the multiplier to use to calculate a 30 percent increase on an amount. In this example, it represents a 30 percent increase over the $10 price that you sell to your distributor. In the next example, 1.6 represents a 40 percent increase.)

If your distributor wants a 40 percent margin, multiply your $10 by 1.67: $10 x 1.67 = $16.70.

This would then be the distributor's product price.

OTHER FACTORS TO FIGURE: How much can your product be sold for in the stores? Although there are always exceptions to this rule, retail price is usually double the wholesale price. That means, for example, that a wholesale price of $16.70 would translate into a retail price of about $33.40.

You say people will buy your product at $40 retail? Then you are looking pretty good. You can raise your price to your distributor, and the distributor can still make the required profit. Or you can leave the price as is, and talk to your distributor about using the "extra" profit for specials, discounts, or advertising. But if your product won't sell for more than $25 retail, you need to go back to the drawing board.

You really need to know your costs and examine what your overhead will be before you talk to a distributor about price (see chapter five for more about figuring costs and overhead). The fewer of these costs you have, the better position you will be in when working with a distributor.

Exclusive or Non-Exclusive Rights

Most distributors will want *exclusive rights* to distribute your product. This means it—not you, not another distributor, not a sales rep, not anyone else—is the only one that is allowed to sell your product for the life of the distribution agreement. However, this exclusivity is not always as limiting as it may at first seem, because many times a distributor will agree to the exclusive right to sell a product only in the market area it serves, leaving the other markets open for other distributors you may wish to seek out.

Most products can be marketed through more than one chain of distribution.

Say your product is a toy for children to use in swimming pools, and you get a distributor that handles toys for children. The distributor sells to stores and catalogs, attends all the trade shows, and advertises to consumers. But you want to try to sell your product to hotels as well, so that guests can use it in the hotels' swimming pools. If your distributor doesn't regularly sell to hotels, then you'll want to limit your exclusive agreement with your distributor to retail outlets and catalogs so that you're free to get another distributor for the hotel market; or sell them yourself direct.

Or perhaps your product can also be used by physical therapists for rehabilitating patients in pool therapy. You may want to get another distributor that sells medical equipment.

LIMIT EXCLUSIVITY TO SPECIFIC MARKET AREAS, WHENEVER POSSIBLE. I would be very reluctant to grant anyone total exclusivity for any product. Most products can be marketed through more than one chain of distribution. It's certainly safe to grant exclusive rights to distributors for their specialized industry, but not for all possible sales.

I have a product that is sold through my company, It's Pawsible that is called a pet caddy. It is a hanging bag designed to store all your pet's accessories in one convenient place. I have exclusive distribution agreements with pet product distributors. But I also have an exclusive distribution agreement with a camping equipment supplier for the same product. Even though both agreements are "exclusive," neither one interferes with the other's intent.

What distributors don't want is for you to show up with a booth of your own at a trade show, competing with them for business. At that point it's simply not worth it for them to have your product in their line.

It's safe to grant exclusive rights to distributors for their specialized industry, but not for all possible sales.

TURN OVER EXISTING ACCOUNTS TO DISTRIBUTORS. If you do get some accounts on your own and then decide to get a distributor, give the distributor those accounts. I know it's hard to give up an account you worked for, especially since the distributor didn't get it for you and you will be making less money on that account from now on. But you must make a commitment to any distributor you hire. The hard feelings you will create by keeping the account to yourself are not worth the profit you'll relinquish to your distributor. You don't need your distributor or your distributor's sales reps harboring resentment because they're losing out on sales that they feel should be theirs.

Distribution Agreements

Before you begin working with a distributor, you should detail your expectations in a *distribution agreement.* This should cover specific accounts, advertising, trade shows, catalogs and sales materials, public relations, geographic market areas covered, and quantities the distributor expects to sell.

Everyone works a little differently, which is why it's best to have all this in writing before you begin. Your distributor may attend six trade shows per year and ask manufacturers (meaning you) to help with expenses for the two biggest. That's okay, as long as you know this and agree to it up front. And you want to know what other costs will be involved. If the distributor does a special sales catalog during the holiday season, will your product be included? Does the distributor do exporting? If so, to what countries, and at what cost? If the distributor sells to a large account at a discount, and therefore needs a greater discount from you, how much will it be? And for what quantities? Find out about incentives for sales reps, bonuses on sales, and monthly "featured" products.

Let the distributor know you want to work together to sell your product, but you *don't* want to be surprised. You don't want, for example, to give the distributor your rock-bottom price and then be asked for further discounts in the future.

Get everything in writing before you begin working with a distributor.

Your agreement should allow sufficient time for the distributor to market your product and realize a profit before you lose patience. Remember, distributors spend money to sell your product, just like you did. You must give them ample time to test its true marketability.

I have included in the Library of Resources an example of a distribution agreement that can serve as a reference when you begin asking questions and selecting a distributor.

Smaller Can Be Better

When I looked for a distributor, I went to the biggest one I could find. I wanted big sales, of course.

But one visit convinced me, once again, that biggest is not always best. This company had a line of some 300 products. All that meant to me was that my product would most likely get lost in the shuffle.

And when the head honcho I met with told me all the glorious things he had done for other products, and all the things he *might* be able to do with mine, I was still not impressed. I wanted someone to recognize my product as a great idea and a great product, as something special.

The second company I went to was much smaller, less pretentious. Could it offer me the same distribution as the big one?

No. But when the head honcho told me how excited he was to be able to include my product in the company's line, it was music to my ears. This company felt right. And it was—for me. The other company may have achieved better sales, but with this company I felt completely comfortable, and it allowed me to remain as involved with my product as I wanted to be. My product became a featured item instead of extra baggage.

Working Directly with Sales Reps

In many cases, the best way you can reach buyers is not by marketing your product through a distributor, but by hiring *sales representatives* to sell it to retail outlets for you. Sales reps make their living from what they sell. They are paid a *commission*, by you, on their sales. Reps charge a commission according to the product and the industry in which they work. Commissions generally range from 3 percent up to 20 percent of the wholesale price. Sometimes, but not often, they go as high as 40 percent, but this is usually when reps sell directly to consumers at retail prices.

To determine the standard commission rates for your industry, you'll need to do some research. Most trade publications and organizations have listings of sales reps who specialize in your industry. Give one a call and explain what your product is. Most will be happy to give you a few minutes of their time to explain the commission structures and so forth.

(text continued on page 129)

A Lesson Learned: Carefully Check Out People Who Offer Help

Getting Little Shirt Anchors into JCPenney was something I dreamed about. I knew there were others out there with products like mine and whoever got into a major store first would be the one whose product the public would ask for. And if I could get into Penney's, the other majors would want to place Little Shirt Anchors in their stores.

Although that's not what happened I want to tell you the story of how I got the Penney's account. It's a real heartbreaker.

I had been in business about one and a half years. I had heard so many lines from people about how they were going to help me get rich quick that, I was just sick of it. I was beginning to realize that the only way to make this work was with hard work and incredible endurance.

Then I got a call one day from a man, I'll call Roger. He asksed me if I would like to sell my product to one of the biggest retailers in the country. Trick question, right? I said yes. I then asked him who was the big account. He wouldn't tell me. In fact he was angry that I wanted to know the account name and implied that I didn't trust him. I was immediately suspicious; he thought that I was being ridiculous and told me so. We really didn't get anywhere with that call.

A few days passed; Roger called back. This man was arrogant and rude to the point of being unbelievable, but his talk of this "big" account started to intrigue me. I have always had a hard and fast policy of only doing business with people when it feels right. If I can't communicate, or I'm in a situation where I feel like I need someone too much, I tend to back away.

Roger, however, reeled me in. He asked me detailed questions about how the Little Shirt Anchor is made, where it is made, had I ever considered changing the design? Questions that made me very uncomfortable. He finally did tell me that he was currently manufacturing something, although not in the juvenile field, and that this account had been buying from him for some time. He explained to me that because he was already a vendor, it would make the sale much easier. I still felt like something was not quite right, though.

(continued on next page)

I wanted to know the name of the account and, fortunately, Roger made it easy to figure out. He started getting careless about things he said, things like his product had been tested in a city near his office, that they had a large catalog, etc. Through these pieces of information, along with others, I figured out that it had to be JCPenney.

This was the account I had always dreamed of. This was the account I had written a letter to several months earlier, only to be rejected.

After much soul searching and discussion with my family, I finally decided to work with Roger, to let him help me get the sale. I gave him a list of happy customers, I told him all the selling points, I told him all the research statistics so that he would be fully armed with information when the buyer asked him, "well, what if this happens?," or "what about that?" I told him every piece of information I could think of to help him sell my product.

All along, of course, he still kept thinking I was in the dark about who the account was, when in fact I had found out what his entire product line was, his home address, even his wife's name.

Roger had his son fly to meet with me, and to pick up a sample of my point-of-purchase (POP) display (see chapter eight for more on this) to take to the meeting. I literally had to meet him at the airport, as he was flying back one hour after he got in. Well, I dressed to kill. I was hoping I would take him off guard, and I did. I confirmed the fact that it was Penney's; I also found out the buyer's name. His son made me swear I wouldn't tell his dad I knew. I agreed.

When the big day came he called me immediately from a pay phone in New York City; that's where the Penney's offices were at that time. He said, "I suppose I can tell you now the account is JCPenney."

"Oh, my," I said with all the surprise I could muster. How did it go? He told me that he had to meet with an assistant buyer, which aggravated him very much. I was hoping he had not treated her as he had treated me in our first few phone conversations. He said he would call me in a couple of weeks to let me know how it went; they hadn't made any decision yet.

(continued on next page)

Two weeks went by, then three. After the fourth week I began to worry. Finally I called him. I asked him if he had any news. "I'm sorry, they didn't like your version of the product." I was confused. "What do you mean, my *version of the product?" He told me that he felt the Little Shirt Anchor was too expensive, so he offered them my product, as well as his cheaper version. Thanks to the success of my product, they were sold on the idea, but they weren't going to buy it; they wanted to buy Roger's, because it was cheaper. He finished with: "Sorry honey, there are other fish in the ocean." Then he hung up.*

To say I was stunned would be putting it mildly. Shocked is more like it. I couldn't believe he had done such a thing. The next three days seemed like an eternity, until I decided the only thing for me to do was to call Penney's. Now, I did not want to put the buyer in the middle of a disagreement between Roger and me; that would not be appropriate. I did feel, however, that it would be okay for me to contact the original buyer I had gotten the rejection from. I wanted to see if I could do a last-minute "save," if there was anything *I could do to get Penney's to reconsider. I would play dumb, as if all I knew was that my sales rep had made a presentation and I was still waiting for word.*

I made up a story about needing to verify the price-per-unit that my rep had quoted her in his presentation. I was going to tell her there was some confusion, and since my rep was out of town, I thought I'd check with her. I didn't really expect to accomplish anything by doing this, other than making myself feel better.

When I called, I asked for that original buyer. The person at the other end of the phone said she was not available. Then she identified herself and asked if she could help me. Imagine my surprise to hear that I was talking to the very same assistant buyer that Roger had met with.

I hesitated only a moment, then, figuring this had to be some kind of omen, I spilled my guts. I mean, the account was lost already, so what difference did it make what she thought of me?

(continued on next page)

She let me finish, saying little, then assured me they would check into the matter and get back to me. When I hung up I felt so much better. I knew that a company the size of Penney's would hardly be concerned about my situation, but it was nice of her to let me get it off my chest. I knew that the "getting back to me" part was just a way to get me off the phone, but that didn't matter. Do you know what the odds are of calling the buying offices at a major chain like Penney's, and reaching the exact same buyer that Roger had spoken with? They have dozens of buyers, and usually people like me are told, "leave a message; she'll call you back."

Sure. So the fact that I had spoken with not only any assistant buyer, but Roger's assistant buyer made me feel that maybe all was not lost.

Nevertheless, life went on for about five months or so. Then one day it happened. My phone rang. The ring sounded like it always did, but on the other end of the line was that very same buyer from Penney's. She told me they had researched my story, checked the information I sent, and concluded that I was the originator of the product. They would like very much to make me a vendor and put my Little Shirt Anchors in their stores.

After that day my life was forever changed. I have great faith in listening to your gut instinct, and I have great faith that sometimes the little guy does come out on top. Thank you JCPenney. And thank you, then assistant buyer.

What's the lessons to be learned from this little tale of mine?

Don't give up easily. Keep at it until you are certain that there's no chance left. Plain, old-fashioned gumption can get you surprisingly far in this business. Also, use common sense and protect yourself. Be sure you have an agreement with your rep that clearly states that he or she cannot handle any other product that might conflict with yours, including something from his own line, and check his or her references. That, of course, is what I should have done.

Finding a Reputable Rep

When I look for a rep for any product, I go to the horse's mouth and ask stores for names of reps they like, reps who service them well, reps they trust. I find that some stores are reluctant to say anything about any reps, but others are happy to share their opinions.

If asking such questions doesn't work for you, or if you are uncomfortable doing it, go back to the library for those listings of trade organizations. These organizations will be happy to help you find sales reps for your particular product. Also check the trade magazines. You will usually find ads from sales reps looking for lines.

How Reps Work

Keeping in mind that sales reps make money from what they sell, you need to realize just where you stand in the marketing chain. Independent reps can handle anywhere from 3 to 20 product lines at a time; you will be only one of these.

Reps can handle anywhere from 3 to 20 product lines at a time; you will be only one of these.

During an appointment with a buyer, reps start by discussing the line they're most likely to get an order on. By the time they get to your product, usually at the end of the presentation, the buyer might be rushing to another meeting. Or maybe the buyer is simply tired of looking at reps' faces all day. Or the reps have used all their best adjectives (exciting, unique, innovative, one-of-a-kind, guaranteed seller . . .) to present those products that come before yours.

Knowing this, you can see why some of the reps' sales punch is gone when they say they've got one more really great item to show.

If their interest is piqued, buyers will often ask, "Who makes that?" If it's a new company they have never heard of, they may be a bit reluctant to place an order. They may say, "Thanks for showing me that. When my customers ask for it, I'll know where to get it." (Reactions like that hold little promise, I'm afraid. After all, how will the customers know to ask for a product if they have never seen it?)

Territory

When you hire reps, you usually need to assign them a territory—a geographic area in which they have the exclusive right to sell. That means you won't hire another rep for that area, and you won't go there to get accounts yourself.

I make sure my reps have every account in their territory. Even if they are accounts that we handle *in-house* or call directly for orders, I still pay my reps a commission on the sale (although it's usually less than they would get on their other accounts). I know we have a commission figured into our price anyway, and this policy makes the rep feel better about me and my product line and my company, which makes them market my product better.

Some companies, on the other hand, keep in-house accounts to themselves and don't give the reps a penny in commission. How you work it is completely up to you.

The Rep Agreement

In the Library of Resources you will find a sample rep agreement you can adapt to your own needs.

I pay my reps on the 25th of each month for orders shipped the previous month. You want to pay based on orders *shipped*, not just *placed*, because you may have to do credit checking or an order may get canceled before it's shipped. If any items are returned, the commission is deducted from the rep's commission the following month. (I'll go into this further in chapter ten.)

How much time should you give reps to prove themselves? That's a debatable point. You don't want to be locked up with reps who aren't working their territory, but you want to give them sufficient time to get going. It probably takes three to six months to really make inroads with a product. If your rep doesn't get anywhere in that length of time, it might be best to make other arrangements.

Motivating Reps

Keep your reps motivated. I send out a letter every couple of weeks to my reps, letting them know about any new developments related to my product, such as publicity (send copies of all TV, newspaper, and magazine write-ups) or positive feedback from customers.

INCENTIVES. You may want to throw in an extra incentive now and then, like a bonus for the rep who sells the most in the month of August or the rep who shows the largest increase in margin of sales. I prefer offering incentives based on margin increases rather than sales increases, because reps may have very different size territories.

Send them birthday cards and get-well cards. Send them a little something during the holidays. Let them know how important they are to you; believe me, it will make a difference when they present your product and speak glowingly about the company that manufactures it.

SELLING DIRECT TO CUSTOMERS

There are different ways to sell directly to customers. For example, you can buy a mailing list of customer names and send out a flyer for your product along with an order form (called *direct mail*), or you might buy a TV or radio commercial or run an ad in a magazine that encourages people to call your 800 number or write in with their orders. You would not sell directly to the customer if you sold through retail outlets via a rep or distributors. You would then be competing with them and that's not something you want to do—there are exceptions to this rule, though few and far between.

Let reps know how important they are to you, and it will make a difference when they present your product.

Despite its variations, selling direct is costly. TV and radio time is expensive. The mailing pieces are expensive, postage is expensive, and mailing lists are expensive to buy. Most successful mail-order sales occur when you have one or two excellent products, get responses for those products, then follow up with a catalog or brochure showing other products the customer can buy. This way you build a list of names of potential customers who are likely to buy more products from you—*if* they liked the first product and *if* they received it quickly.

I know of a great many inventors who have run ads in magazines, newspapers, or newsletters, only to find that, at best, the resulting sales barely covered the cost of the ads themselves. Yet another obstacle for a single-product manufacturer: It's difficult to support a sales effort with one product. You need to do some careful research before deciding whether direct mail is right for you.

Is It Going to Work for You?

Look carefully at publications that run ads for products similar to yours. Call and ask for back issues and check to see how many mail-order ads, approximately the same size as the one you're considering, each issue contains. Do they continue running from issue to issue, or are there a lot of one-, two-, or three-time advertisers who then disappear? That's a clue the ads didn't perform.

One aspect of direct mail never varies: It is costly.

CALL UP OTHERS WHO'VE PLACED SIMILAR ADS. Once you've found ads that have been running for several months, choose some that look appealing and that advertise a practical product, and call the advertisers. (Make sure they're ads for products that don't compete with yours.) Tell them you're just starting out and you'd like to talk to someone with experience in advertising. Make sure you mention experience. It's very flattering, and it makes people feel almost obligated to answer your questions.

Tell them right off what your product is, so they'll feel comfortable knowing you're not a competitor trying to get information. Ask if

Getting Authorization to Accept Credit Cards

You can set yourself up to accept credit cards as payment for goods by contacting your bank. Tell them you'd like to set up a merchant credit card account. You will need to fill out an application on your company and wait for approval.

their ads have been successful in the publication. Don't ask about the number of sales or anything else so specific.

If the information you get sounds encouraging, you may want to give direct-mail advertising a try. But first read chapter eight for information on how to create direct-mail ads. You may also want to read one or more of these books: *The Golden Mailbox*, Ted Nichols (Dearborn Financial Publications, 1993); *The Complete Direct Marketing Sourcebook*, John Cremer (John Wiley & Sons, 1992); *Secrets of Successful Direct Mail*, Richard Benson (NTC Publishing Group, 1991); *New Direct Marketing*, David Shepard Associates (Irwin Press, 1990/1995).

DIRECT MAIL: KNOCKING ON YOUR PROSPECT'S DOOR

Direct mail—mailing a promotion piece directly to your customer, at home, at the office, or wherever—is a very popular way of promoting a product or service, as the stack of stuff in your mailbox proves. Obviously it's working for someone. You should do some careful research to decide whether it can work for you.

Just as with print ads in magazines, direct mail allows you to target either the trade or the consumer. If you've got a new dog collar to sell, you might send a brochure out to everyone in Texas who owns a dog. Or you might send a brochure to every pet store in Texas to see if they want to carry your dog collar. Either way, working from computer-generated databases containing millions of names and statistics, you can create selection criteria to help you market to your target customers.

The Risks of Direct Mail

No matter how selective you are with your list and how carefully you narrow the field of prospective customers, it's a gamble, and an expensive one at that. Many people who receive "junk mail" don't even open it up; it goes directly from mail box to wastepaper basket. Don't waste your money on a big mailing right at the start; do a test mailing first—before you invest several hundred dollars in all the printed material and postage you'll need to do a big mailing, or *mass mailing*.

Not many manufacturers are successful with direct mail. After I started running mail-order ads, I determined that I needed to do follow-up. I created a mail-order catalog that went to each customer who purchased a product through the ads, and I was able to make money that way. But I never made any money by advertising just one product in a direct-mail ad.

Direct mail usually works better for local restaurants or service-oriented businesses than for consumer products. Restaurants and cleaning services can send out a coupon for a "buy one, get one free" meal, or offer you 10 percent off a carpet cleaning. Products adapt less well to direct-mail ads because the "touch me, turn me upside down" aspect of buying is absent.

No matter how careful you are in selecting your target audience, direct mail is a gamble.

How to Do It

If you want to send your direct-mail piece directly to customers, you'll need to get your hands on customer mailing lists. Sources of such lists abound. Catalogs, credit card companies, department stores, magazines, health clubs, store chains, lending institutions, and professional associations are just a few gold mines of prospective customer names and addresses.

If you want to further pinpoint your customer, you can turn to a company that specializes in mailing lists. Ask at the reference desk of your library for listings of such companies. One good directory is the *Standard Rate and Data Mailing List Directory*.

Here's how they work: You provide them with the criteria, or demographics, for people or businesses you want to solicit—say, retired people who own an RV and a cat and live on the West Coast—and they will do their best to provide you with a mailing list of just those customers. Demographics can also include occupations, income levels, ages, sex, hobbies, and a myriad of other criteria.

In most cases, mailing-list companies will either provide you with the mailing labels, mail your ads for you and charge you for the postage, or provide you with a computer disk with the names and addresses.

Ad agencies always have access to mailing lists as well, though obtaining them through agencies may cost you more because they add on a service charge. You can also contact sales reps at magazines in your field; they may be willing to help you (a potential future advertiser) get your business off the ground by providing you with their mailing lists. Many magazines, newsletters, and other publications will also sell you their mailing list, or even let you use it free of charge if you are an advertiser.

Before you try direct mailings, think about what gets you to open a piece of junk mail: Is it the headline on the envelope? The color photo? The name of the sender?

Direct-Mail Dos and Don'ts

- **If you obtain a list through a mailing-list company, be sure to ask if you may use the names on the list more than once for the same price.** When companies specifically indicate that you are purchasing the list for one-time use, they will check to make sure you don't try anything funny. They have dummy names on their mailing lists that only they can recognize. If you use the list for repeated mailings, they will know and will charge you accordingly.
- **Always ask ahead if there will be a charge for an optional mailing-list service.** You may be offered several options that increase the cost of your mailing list. For example, if you are buying a list for trophy shops, the mailing-list company may ask if you want to include the name of the owner or manager and charge extra for this.
- **Ask if you will get a refund for names that are no longer valid, such as stores that have moved or gone out of business.**

If there is no refund policy, ask your list salesperson to give you a guarantee that a certain percentage of the names will be valid. It's impossible for these lists to be 100 percent up to date, but if you get 50 percent of your mailings returned to you, undeliverable, you should have some recourse.

• **When you prepare your direct-mail literature, remember that weight is money.** Keep your mailing package under an ounce. Then, if a mailing-list company isn't taking care of the mailing for you, go to your local post office and get a bulk mail permit. You can get information on how to use the bulk mail system through the post office. Some even conduct free classes for people who want to learn how to use a bulk mail permit.

Companies that specialize in mailing lists can pinpoint your target customers—say, retired people who own an RV and a cat who live on the West Coast.

Essentially, doing a bulk mailing means you pay much less than first-class postage because you presort the mail for the post office and the mail is not first-class mail. But bulk mail also takes longer to get to your customer—in many cases, as long as three or four weeks, which is fine if the material is not date sensitive; otherwise, forget it.

Who Pays Sales Tax? Say Rubbermaid buys plastic from a plastic manufacturer. Rubbermaid pays wholesale price for the plastic because the plastic is not the end product and Rubbermaid is the manufacturer, not the end user. The company does not have to pay tax. Rubbermaid turns the plastic into trash cans and sells them to Kmart. Kmart pays wholesale price to Rubbermaid because it is not the end user. Kmart does not pay sales tax, either.

You go to Kmart to buy a Rubbermaid trash can. You buy it at retail because you are the end user. You pay sales tax to Kmart. The sales tax is collected only on the final sale for two reasons: to prevent sales tax from being collected for the same materials over and over (the plastic for the trash cans, for instance, has been purchased three times) and so that the tax can be collected on the biggest sale—the retail sale—which means that the government will collect the biggest tax from this transaction.

If you sell your product to your local store, you will sell it at wholesale and not worry about collecting taxes. If you sell your product at a fair, a swap meet, or through a mail-order ad, you will sell it retail, to the end user, which means you will have to collect the sales tax (if applicable in that state). When you apply for your business license with your city, you will be notified if you are required to collect sales tax.

MAIL-ORDER CATALOGS

The mail-order catalog business has been growing by leaps and bounds in the past ten years. There are catalogs that sell everything from gardening supplies to 100 percent cotton clothing. All you need to do is order once and it will seem like your name is turning up on every catalog list that exists.

Catalogs are terrific outlets for a single-product manufacturer because they pride themselves on offering products that consumers cannot find in stores.

A Great Showcase for Your Product

My mailbox is full of catalogs every day. I find these publications delightful, not only because I buy things in them, but because they are terrific outlets for a single-product manufacturer. Catalogs pride themselves on offering products that consumers cannot find in stores.

For this reason, you have an excellent chance of getting your product placed in a catalog—no matter what the size of your company, no matter how new the product. In fact, the newer the better, in most cases. Catalogs don't wait for customers to ask for a product. They educate their customers about what's out there.

The Ins and Outs of Catalog Placement

Working with catalogs is a bit tricky. Most will want a discount on your product, which in many cases is justified. They want to be able to offer a good price even after the customer has paid shipping and handling costs. So figure out what discounts you can offer before you approach them. (See Pricing Your Product for Distributors, earlier in this chapter, for some guidelines.)

The other issue is delivery, or *fulfillment.* You absolutely must be able to deliver your product on time. If you are late, you will either be charged a late fee or dropped from the catalog—no matter how well the product sells. If the catalog has to hold up an order for six products because yours isn't in stock, you are doomed. And if they ship the other five products, then have to send yours separately, they have to pay an extra shipping cost they did not charge their customer for. That is mail-order death.

Catalogs pride themselves on offering products that consumers cannot find in stores.

Be well aware: When your product is advertised in a catalog or mail-order advertisement, be prepared to fill orders quickly! I know of one company that got its product into the catalog of a mail-order giant. The company's items were custom made, so, obviously, they had to be produced after the orders came in. The employees got those orders filled within 48 hours, no matter what. They worked all night long, they worked with colds, they worked on holidays, they worked their fingers off.

But it all paid off. The mail-order business was very impressed because their customers received the product within a week of ordering, and three out of ten ordered a second one. (An incredible response.) The quality was there, and so was the service. The company and the mail-order giant now have a multimillion-dollar business with dozens of custom-made products. And they still fill those orders within 48 hours!

(According to the Post Master General in California, the required length of time to fill an order is between the company and the customer. If, however, the customer becomes dissatisfied with any part of the service—including delivery times—and a complaint is filed with the post office, the company's policies and handling of that particular customer as well as other aspects of its business may be investigated.)

Most catalogs will give you projections of what they expect to sell, but these estimates are bound to be off. So don't sit back and wait for the next order to be placed, which the catalog said would come next month. It may come next week, with "Cancel if not sent by ________" in big letters at the top. You don't want to lose an order.

Submitting Your Product to Catalogs

If you'd like to get your product into a catalog, call the catalog company's customer service and ask for submission procedures. If you need to find catalogs appropriate for your product, check your local library; there are books that contain listings of nothing but mail-order catalogs, listed by category. *Don't submit* your product to any catalog before you've seen that catalog. You may find, for instance, that a catalog called *Bird In the Hand* doesn't have anything to do with birds, but carries books on positive thinking. Once again, doing your homework benefits you tremendously.

Two tricky areas when it comes to catalog placement: discounts and delivery.

Don't be afraid to send your product to a catalog that already carries something similar. The catalog may be unhappy with the current supplier and may be willing to give you an opportunity. I have knocked products out of catalogs many times; not out of spite or vengeance, but because business is business. I know what the catalog needs and I provide it, plain and simple.

There are many different arrangements under which you may work with a catalog. The most common is to sell them products, just as you would to a retail store.

Some catalogs may request that you *drop ship* a product directly to the customer who ordered it. This is usually done on personalized items, as it saves time, instead of shipping an item to the catalog just to have them turn around and ship it to the customer.

Don't be afraid to send your product to a catalog that already carries something similar.

Occasionally a catalog will want to buy items on consignment. That is, they may order 500 from you, pay you as they sell them, then return the unsold portion after the catalog has run its life, which is often about three to five months.

Many catalog companies will ask for a *catalog discount* on their orders; they often base this on the fact that you can use less-expensive packaging for their products. For example, a toy that needs a brightly colored, eye-catching box to be sold in a store, may only need a plain box to be shipped to a catalog customer. The catalog customer has already made her decision to purchase, she doesn't need the sales effort of the packaging.

Or you may be able to send a product with no packaging at all. This will be determined by the catalog company at the time it orders the product from you.

CHAPTER EIGHT

ADVERTISING AND PROMOTION

From Brochures to Big Time

With the use of compelling words and enticing colors and images, even the most uninteresting products can become appealing. A form of brainwashing? Perhaps. But whatever it is, if done correctly, the advertisement makes you feel like you couldn't—or shouldn't—possibly do without it.

CREATING DEMAND FOR YOUR PRODUCT

When a major company launches a new product, it can create demand for the product even before it's on the market. This is accomplished primarily through advertising and promotion. It's done not only for products, but for services, food items, and even movies.

The Lion King and *Batman* lunch boxes, baseball caps, T-shirts, and toys were on the shelves long before the movies were released. Clever marketing departments used these products to create heavy consumer demand for the movies. But creating this kind of demand requires a lot of money, a creative marketing team, and a strong track record of past successes that customers remember.

Convince consumers that your product will make them richer, stronger, skinnier, or simply happier, and you're on your way.

How can you, the lone inventor, possibly compete in a world that's full of more "show-me" customers than customers willing to give the unknown a try? Well, if you convince consumers that your product will make them richer, stronger, skinnier, smarter, more attractive, more popular, more efficient, or simply happier, you're on your way.

The rapid growth of television shopping channels, mail-order catalogs, and infomercials enables entrepreneurs like you to reach consumers right at home, where you can show customers your product and sell it to them at the same time.

Can there be any question about whether to advertise? You already know that advertising a product means more people will see it, which means more people will be likely to buy it. But before you plunge in, there are two important issues to consider: the cost of advertising and the placement of your ads—which lead me first into a discussion of working with advertising professionals.

HIRING AD AGENCIES

Ad agencies can do as little as design your business cards or as much as launch a nationwide advertising campaign. An agency may design everything from your logo to your 60-second commercial, manage all placement for magazine, newspaper, TV, and radio ads, and handle all your mailings—including licking the stamps.

Because they're professionals, ad agencies will generally produce a campaign that's superior in quality to that of an amateur. But you'll pay a premium for their services.

Is it worth it?

That's up to you. Have you ever worked in advertising? Do you have any experience in writing copy for an ad or laying out an ad? Do you have a friend or relative who has worked in printing or photography and could assist you?

If the answer to all the above is no, then you would be wise to at least consult with an ad agency before beginning. If you are going to advertise, it is worth doing it right.

Ad agencies will generally produce a superior campaign—but you'll pay a premium for their services.

AD PLACEMENT: TRADE VS. CONSUMER PUBLICATIONS

There are two distinct types of written publications: *trade* and *consumer*. Both can take the form of magazines, newsletters, journals, or books. But trade publications are targeted to a specific trade or industry—say, swimming pool manufacturers—while consumer publications are targeted to, yes, consumers—people like us.

Consumer publications and trade publications each reach very different target markets. So an advertisement for the same product in each publication should aim to accomplish two entirely different things. Your ad in a consumer publication might emphasize your product's convenience and durability. Your ad in a trade publication might emphasize the product's effective packaging, which virtually sells the product itself, or announce that buyers will receive discounts on orders of 100 or more.

Consumers' concerns are very different from those of the trade magazine reader, so you need to determine which buyer—the consumer or the distributor and retailer—you want to target before deciding where to advertise.

To reach this decision, answer these questions:

- Do you want to fill individual orders from your home?
- Are you prepared to handle a lot of calls and paperwork?

If your answer is yes to both, then your ad belongs in a consumer publication.

You need to determine which buyer you want to target before deciding where to advertise.

On the other hand, if you've decided to use sales reps or a distributor to encourage stores to buy your product that they can then sell to consumers, then your ad belongs in a trade publication.

Perhaps you want to get your product into stores but want to advertise it to consumers so they go into stores and ask for it. If this is the case, then you should advertise in both types of publications.

Advertising in Publications

If you're interested in running an ad in a magazine or other publication, the first thing you need to do is find out which publications target your customer. Call them and ask a sales representative to send you a media kit, which will include information on the publication's readership and advertising rates.

You will notice that in almost all cases the cost of running a one-time ad is incredibly high. The price usually gets easier to swallow if you're willing to run your ad 12 times—but that means a bill each month for a whole year! A daunting thought, particularly

A Lesson Learned: Be Careful What Tactics You Use to Get Attention

I recently was trying to get a product placed in a major chain store. I had tried for years to get this buyer to look at it, but there was nothing I could do. So I took drastic steps and sent him flowers. Not once but four times. And not once did I sign my name. I figured maybe I needed to get his attention, then talk about the product.

I don't dare mention the product or the buyer's name because I don't want to get in trouble all over again. That's right. When I finally called him to tell him who I was, I expected a laugh or a smile. What I got instead was his anger that I had caused him valuable time wondering about this mystery, when he had better things to wonder about.

So beware: even your best efforts may backfire. I assured him I meant no harm and apologized for causing him any concern.

(Note: Many companies have a very strict policy regarding their buyers receiving gifts of any kind from manufacturers or reps; so don't get them into trouble. Flowers or candy could cost a buyer her job.)

since you have no idea how effective your ad will be. But if you've done your research, you have at least a fair idea of its potential impact.

Negotiating for Ad Space

If, after you've done some investigating, the publication still seems like an appealing place to run an ad, call the sales rep back. Explain that you are interested in running an ad long term. Tell the rep you'll place the ad directly, which means you'll get what they call an agency discount. (If an ad agency places an ad for a client, they receive a discounted rate.) Also tell them it's a direct response ad and you really need a discount for that. Some publications offer this discount routinely; some do not.

Mention that if the ad is successful, you plan to increase the size right away, but you're just starting out and need all the help you can get now. You'll be surprised how far down you can bring the cost of an ad.

Never pay published rates from a rate card. The advertising sales rep usually has some room to negotiate. I ran an ad for our company that had a list price of $33,000. That's right—$33,000 for a one-time, full-page, four-color ad. I got the ad for $17,981. That's still a lot of money, but you must admit it's one heck of a discount.

Tell the sales rep that if you run an ad you expect some editorial space. That means you'd like a little write-up in their new product section or wherever they usually write about new products their readers may be interested in. Most do it, so tell them you want it.

If you run an ad that looks cheap, potential customers will assume your product is cheap as well.

Creating the Ad

Once you've agreed on a price, you need to have the ad produced. Ask the sales rep if the magazine can help you with layout, design, typesetting, etc., for producing your ad, and whether these services can be included in the cost, since you'd like to run your ad so many times. If the publication can do it, great. If not, that's another expense for you.

Whatever you do, don't run an ad that looks cheap. If it looks cheap, you will turn off many potential customers because they will assume your product is cheap as well. You *can* be creative without spending tons of money. Get some help; look in your phone book and call around. See more about how to create an ad later in this chapter.

Aiming Small: Advertising in the Neighborhood

If you're working with a tight budget, you might want to begin by advertising your product in small local newspapers and newsletters. Your local or regional publications have ad rates that are a *lot* easier to swallow than some of the big national magazines. And it's perfectly okay to start small.

Remember, every time you place an ad you are paying for, you should ask the publication if you can get some editorial space for your product as well. That way, you might be listed in the new products column on page 4 and have your ad on page 11. Two exposures in the same publication double your impact on the customer.

Targeting Clubs and Associations

Most associations have newsletters for their members, and there are groups devoted to practically every profession, from CPAs to dentists, and every interest, from surfing to automobile collecting. Then contact them and ask if you can advertise in their newsletters, distribute flyers to members, or even do a presentation at one of their meetings. One source is *The Encyclopedia of Associations*, published annually by Gale. Use the encyclopedia or contact a local organization or even a specialty store for other referrals—watch your newspaper for "special events." Sometimes club meetings are listed.

An endorsement from an association president lends credibility to your product and may encourage other members to try it.

It is not that difficult to get an endorsement. Usually the group wants the publicity from the endorsement as much as you want the endorsement. Send a press kit (chapter 9), product information, and information about yourself.

An endorsement from the president or a member of such a group lends credibility to your product and may encourage other members to try it. Even if the group represents just one section of your potential customer base, it's a worthwhile place to start. You don't have to sell the whole country at once. Be patient: Give your product an opportunity to follow the regular rhythm of growth and acceptance.

Making a Big Splash in a Small Pond

If you advertise locally, be creative with your print ads. Consider offering a limited-time sale, or a coupon, or a premium incentive. A premium incentive is something along the lines of: "Buy three cans of weed killer and get a free sprayer!" Be inventive and different. In advertising, the most unique and original usually wins.

I still remember a local company that did tune-ups on cars. With every tune-up done in the month of March, they offered a gift certificate for a free manicure at a local chain of nail salons. They were flooded with female customers. You might want to try something like this yourself. Even if you don't get sales from everyone who sees the ad, you'll get people to notice you. You'll get them talking about you. They may not be customers today, but they may be next year. If your ad has enough personality, they'll remember you long after the ad is gone.

You want as many people as possible to read your ad, so your *catch phrase* should do just that—catch them.

Creating a Print Ad

You want your ad to fit the publication. If all the ads you see are flashy and creative, with a high-quality look, and you don't feel you can create the same look on your own, then find a suitable publication in which your ad will fit in.

If you decide to write and produce your ad yourself rather than hire an ad agency, you can construct it fairly inexpensively and easily. Where to begin? Think of a way to grab your customers' attention. Many times the best approach is to ask a question. For example, "Are you ready to risk your family's safety?" could be an opening for an ad for burglar alarms for your home. You want as many people as possible to read your ad, so your *catch phrase* should do just that—catch them.

The text, or *copy*, should be short and to the point. The more cluttered an ad looks, the less likely anyone is to read it. Don't assume your product's appeal is enough to make anyone read boring, long-winded copy. Following the opening, state the most important facts in an easy-to-read style (bold letters, few words). You can aim to amuse (perhaps with a well-drawn cartoon), include an endorsement (don't bother with an endorsement from a relatively unknown person and make sure you have written permission from the endorsee), or use an anecdote. The size of the ad will largely determine which method you choose.

For more help with writing ad copy, check your local library for books on the subject. They can give you many examples and guidelines.

Typesetting and Art

Typesetting is a method of preparing all the text to be reprinted in the ad or made into film for the printing of the ad. Sometimes the publication that is going to run your ad will take care of typesetting for you. If not, you can have an independent printer or print shop do it for you. Or you can do it yourself with desktop publishing software. Sophisticated programs and laser printers can turn out very professional-looking type these days.

Be sure to ask the publication for a spec sheet, which is a list of specifications for the size of artwork for the ad.

You'll want to talk to several graphic designers to find the best price for laying out your ad. They'll want to know how much typesetting is involved and how many photographs or illustrations; it's helpful to provide a pencil sketch of the ad as you envision it in your mind. Include your ad copy or an estimate of how many lines of text you expect it to be.

If you are going to use any artwork or photographs in your ad, you need to prepare these according to the requirements of the publication you plan to advertise in. Be sure to ask the publication for a spec sheet, which is a list of specifications that spells out such things as the dimensions of the ad, and if they require film or if they can work with camera ready art. If you are sending film, what kind of line screen do they need? What are the specifications on a full-page or bleed photo? And so on. Most printers will be able to assist you in putting your ad together to suit the requirements of the publication.

Be sure to ask lots of questions when you talk to photographers, artists, printers, and anyone else who works with you on your ad. By listening and learning, you'll be able to get the most "bang for the buck" with your advertising. For example, ask the printer to see paper samples; a cheaper selection may be suitable. Ask a printer about being part of a "gang" or "combo" run to save money. Ask an illustrator if creating an image in a computer rather than on paper will save money on future printings. Most of all, ask if there is a less-expensive way to achieve your desired results.

TV and Radio: Seeing and Hearing Is Believing

Radio or television advertising is a bit more of a challenge. One big reason is production cost. It costs more to put together a 30-second TV commercial of reasonable quality than it does to create several print ads. However, a TV ad gives you that next-best-thing-to-being-there effect, and an opportunity to show your product in use. Thanks to the boom in local cable companies, air time for commercials is becoming more accessible and affordable for individual entrepreneurs; sometimes you can even get the network to foot the bill.

Is Your Product Meant for TV Stardom?

Note all products are candidates for the TV market. Talk to specialists in this field—call your local TV stations and talk to someone in the advertising sales department, or ask the advice of an ad agency that specializes in the television market. Keep in mind, however, that not everyone will be objective: Salespeople may profit by convincing you that your product is "just right" for TV.

Before you let anyone convince you that your product should be advertised on TV, do some homework. It's simple and obvious:

watch a lot of TV, particularly local and cable channels and don't hit the remote when the ads come on—pay attention! "Call 1-800 . . . or send $24.95 plus shipping and handling to the address on your screen." You've seen these plugs before, but watch them with a new eye. See what appeals to you and what doesn't. Get some ideas for pitching your own product.

Start by deciding what you want the TV viewer to do. Call and place an order? Will they call your 800 number and talk to you? Will you hire a service to take 800 number orders? Are you ready to accept charge cards? (See chapter seven for getting authorization to accept credit cards.) These are just a few of the things you must consider before setting up a TV campaign.

The Price Factor

The ad should be run during different times on different days and run for at least one month.

Your product's price is a major determinant of whether TV or radio advertising is for you. If you plan to share the proceeds with others for help with production or air time, you need to be able to double, triple, and sometimes quadruple your pricing. Can your product support this?

Don't answer that question until you talk to the TV marketing experts. You may be surprised. I've seen a 30-minute infomercial for car wax that costs $49 a bottle—which, in a retail store, would be a ludicrous price. But after watching one amazing demonstration after another and hearing the cheers of the studio audience (even though I knew they had incentive to cheer) after every one of these miraculous performances, $49 started sounding like a good price. And I could make three easy payments of only $16.33 each! Who could pass that up?

Get your money's worth; do repeat broadcasts. With TV and radio, you are wasting money if you don't run your commercial often enough. Don't bother to go to the trouble of producing the ad and then run it only three or four times. You have to be financially capable of keeping the ad in front of people. The ad should be run during different times on different days. Run it for at least one month to give yourself some good statistics.

Who Will You Reach?

As with mailing lists, you can, to some extent, pinpoint your television-watching customer. Each program on television can provide you with viewer demographics. You can also be selective about which markets or geographical areas you are most interested in. Maybe you just want to advertise locally, within a 30-mile radius. Or you may just want to target large cities. The same holds true for radio. You can get the same demographic information for listeners that you can for TV viewers. Each show or feature will have a different demographic. Check with the advertising departments of the stations for information.

But it can be costly to reach the kinds of viewers, or the number of viewers, you want. It costs much more to run a commercial during a soap opera on a major network than during the midnight movie on a local channel. The more viewers your ad will reach, the more it will cost you.

Radio advertising is generally less costly than television because the ad is less expensive to produce; however, the air time can be just as expensive, depending on what shows and times you fall into. Just as with TV, prime time is more expensive than the middle of the night. You have fewer customers listening, so you pay less.

Working With TV Stations

The more excited a TV station or ad company is about your product, the more willing it will be to put together an ad program for you. So don't be shy; make people aware that potential sales for your product are terrific. Be prepared to answer questions about your ability to supply the product that will be sold and any other advertising or promotion you have done.

If you want to pay the station or ad company a high percentage of what is sold, you may be able to run your advertisements free of charge. If the percentage is high enough, the station may also pay to produce the commercial. Smaller percentages of sales or no percentage results in higher fees for you all the way around. You will need to ask the ad sales department of each local station or TV production company about what sorts of programs, or advertising arrangements, they offer, including the financial investment and percentages for each. Arrangements can vary from 1) no up-front costs for you for creating the ad or running the ad—then "paying" the production company/station up to 60 percent of the revenue; 2) a smaller amount paid up front for ad production and no fee for the air time—then paying the station 20 percent; or 3) you could pay all production and airtime and you would get 100 percent of sales.

Radio and television are powerful and expensive media; don't buy into them until you've done diligent research.

I just did a video/commercial deal in which we paid approximately $3,500 for production of a 60-second commercial, then paid 10 percent of revenues generated from sales to the production company instead of paying air time. We do not get to select what time of day or day of the week our commercial will air, or even during what programs they will appear. We just select large metropolitan markets and see what we get.

Again, many ad agencies that work in television can also assist you in locating a production company that can put together a package deal to meet your financial capabilities. They put products and companies together with television channels all the time.

Along with specific channel packages, there are independent production companies that specialize in putting a commercial together and then coordinate placement and even order taking. You can find these in your phone book under "Video" or "Television" or "Motion Picture" production. Ask for references before working with independent companies that offer to "put together a package" for you.

Learn as much as you can about your area, your customers, and audiences for the shows on which you're considering advertising. Radio and television are powerful and expensive media; don't buy into them until you've done diligent research. For many people this type of advertising has been the key to success. It might also be yours.

GOING DUTCH: COOPERATIVE ADVERTISING

Perhaps you've noticed that retail store ads, usually in the form of flyers, store newsletters, or magazine spreads, sometimes include brand names of products, such as John Deere mowers or Purina Dog Chow. Such ads are often the result of cooperative advertising between manufacturers and retailers.

Chances are your product will be featured with name-brand products, which will lend it credibility.

A retailer planning to run a six-page ad may approach manufacturers and offer them a deal: If they'd like to contribute to the cost of the ad, either directly or by offering a discount to the store on their product, the store will feature their product(s) in the ad. The retailer will usually include companies' names and logos for a little extra publicity.

Small-Scale Co-op Ads

Larger chain stores make such co-op advertising deals at the cost of thousands, sometimes millions, of dollars to manufacturers. But you can also strike up such arrangements on a smaller level, through local stores or small chains of specialty stores. The key to this type of advertising is to let your retail accounts know you are interested in participating. If you don't approach them, they may assume you are not interested in or financially able to make such a deal.

Co-op advertising offers several advantages:

- There is no cost to you to produce the ad.

• The customer reading the ad will know immediately where to purchase the product, and the ad will help link your product to a particular retail establishment.

• Chances are your product will be featured with name-brand products, which lends it credibility.

It sounds great, doesn't it? It can be, but again, you must do your homework. Ask stores to show you other such ads they have run. Make sure you have a clear understanding of exactly what you're paying for—how much space your product will get, whether any competing products will be in the ad, etc. Be sure you're not going to foot the bill for an entire page and then end up with a tiny corner.

PRODUCT PROMOTION WITH POINT-OF-PURCHASE DISPLAYS

Point-of-purchase displays, or *POPs*, are placed in retail stores to help display and sell a product. In convenience stores you may have noticed POPs on the counter: colorful cardboard boxes or small multi-tiered shelving displaying key chains, baseball cards, or some other small item that's often thought of as an impulse buy. Grocery stores often have freestanding POP displays of aspirin, pocket flashlights, eyeglass repair kits, and the like at the checkout counters.

Countertop POPs

POPs come in many shapes and sizes; your product will determine which type, if any, is best for you. They can really get your product out there in the public eye, but getting them out there to begin with can be a real challenge. POPs are not cheap to manufacture, and many stores are reluctant to use them for all but their biggest-selling products because of the space they take up, on the counter or on the store floor.

A countertop display is probably your least expensive option. A small product may lend itself very nicely to a countertop display, preferably placed somewhere near the cash register in the store. In such a location, customers will be exposed to the product for a few moments—or minutes, depending on how speedy the cashier or clerk is—as they wait to pay for a purchase or ask a question. The more the exposure, the more the sales, both for you and the store, or so they say in retailing.

A counter display can take the form of a *dump*, a freestanding, boxlike display into which the product is dumped. This sort of display works well for odd-size products such as yo-yos or sprinkler heads. A counter display could also be used for a product that either stacks or hangs.

Counter displays are a good way to expose your product to customers as they wait to pay for a purchase or ask a question.

Larger POPs

The costs of larger POPs can be considerable. For single-product manufacturers, there's another obstacle to using them as well: Freestanding floor displays, because of their size, are much harder to place in stores, unless your product really takes off. Specialty stores don't use freestanding displays as often as larger chain stores do; they simply don't have the space. And specialty stores are more likely to be your customers when you get started because of the small size of your company, your lack of a track record, and your product's uniqueness.

All in all, floor displays are probably not the best way to go when you're starting out. Wait and see how your sales do, then see if the stores seem interested in one. After you have established a relationship with store buyers, they can help you make such decisions.

Display Basics

POP displays are custom made to fit your product. Dump-type POPs are usually made of a very durable material such as heavy cardboard or plastic that can stand up to repeated use. POPs may also take the form of small wire trees on which products are hung with a header card or sign at the top to describe them. Very low-priced items may come packaged in their own POP display—a lightweight cardboard box that, when opened and folded on scored lines, creates a stand-up header card and makes the product accessible to the customer. This type of display is meant to be thrown out when empty.

A counter POP must be of reasonably small size and weight, but large enough to be noticed. The POP display I used for the Little Shirt Anchor had a footprint (the amount of counter or floor space a display takes up) of 10 by 10 inches. It consisted of a tray on which to stack the product and a header card that stood 15 inches high in back. Many people told me it was too big and stores wouldn't use it. But I wanted a display that would hold at least 18 pieces so that the stores would buy a good quantity at a time. Fortunately, the majority of my specialty stores not only used it but loved it.

Floor displays are probably not the best way to go when you're starting out.

Whether you choose a counter display or a free-standing display, you will want to make sure it is *simple* to assemble. Don't count on store employees to look at directions for more than 30 seconds or play around with it for more than a few minutes. Here again, it's a good idea to turn to an expert for advice. Try contacting display manufacturers or ad agencies listed in your local phone book. Some printers may also be able to offer advice.

POP Problems

As I mentioned earlier, the main problem with using POPs is that there is often very limited store space available for them. Many products have POP displays that store owners don't use; they simply take up too much space in comparison to the revenue they generate.

If your product proves to be very popular, store owners may ask you if you have a POP display they can use. Otherwise, you may have to talk them into using it. Either way, you want to make sure the display is as effective as possible and takes up as little space as possible.

As I also mentioned, POPs aren't cheap. Can you charge the stores for these displays to help recoup some of the cost? Usually not. Customarily, you offer stores a display if they buy enough products to fill it. If your display holds 48 pieces and your average order is 24 pieces from your stores, the display will pay for itself in no time, because the stores will be ordering twice as many items.

That's why I wanted my Little Shirt Anchor display to hold 18 pieces. My average order from smaller specialty stores was 12 pieces. I gave them the display if they ordered 18 pieces; but because they usually bought by the dozen, my average order went up to 24 pieces. They kept the rest on hand to fill the display when it was low. One side note—I also learned that 18 was a mistake, because it left the store with 6 extra pieces in a shipment of 24.

TRADE SHOWS

If you've ever been to a trade show, you'll understand why it should be called a circus. The preparation, the handshaking and smiling all day, and the relentlessly fast pace and high energy really take the wind out of you. In chapter two you'll find information about locating trade shows, how to prepare for them, what to expect, etc. Here I'll explore using trade shows to advertise your product.

Trade shows should be considered part of your advertising budget. They're a great opportunity to give a large percentage of your potential buyers an up-close-and-personal look at what your product has to offer.

Grabbing Attention for Your Product

Your imagination and your budget are the only limits when it comes to advertising your product at a trade show. You might decide to offer product demonstrations. You might give away free samples of your product. The only rule is the one that applies to all advertising:

You must stand out from the crowd. Be as different, exciting, and original as possible—but always maintain a professional demeanor (even if those around you don't).

At trade shows you must stand out from the crowd.

This is a chance to show not only your product but your company's personality. Here are some of the things I have seen at various trade shows to attract attention (keep in mind that these attractions won't go over equally well at all shows):

- An attractive man and woman dressed in skin-tight outfits demonstrating country line dancing
- A baby grand piano and a pianist dressed in a black tuxedo
- Custom-made candies in the shape of a company's logo
- A man taking a bubble bath
- Thin blond women in short shorts and high heels holding baby bottles
- Tall, thin, blond women in bikinis holding crescent wrenches
- Celebrities offering autographs
- A five-foot iguana that you could hold and with which you could have your picture taken
- Foot massages

Of course, having a great attraction at your booth doesn't guarantee you're going to sell more. No amount of giveaways and distractions will coerce a buyer to make a purchase they do not want to make. What you will be doing is getting those buyers to look at your product, meet you, and get a feeling for your company. Maybe this product is not for them, but your next one will be. Won't you be happy that you worked on establishing a relationship with customers at this early stage in your company's life?

Trade Show Tips

GET THE NAMES AND ADDRESSES OF ALL WHO STOP BY. Be sure to have a handy place for buyers to throw their business cards so they won't be burdened with carrying your product information around the show all day long (and so you have a record of who stopped by). A conveniently placed fishbowl will suffice. You might also provide a sign-in sheet for those who don't have a card.

Check show restrictions. Remember to check with trade show organizers to make sure your display is okay and you're not breaking any fire codes or leash laws. Also, keep in mind that most trade shows do not allow drawings or contests of any kind in individual showrooms. Be sure to check with your show organizers if you want to offer a free case of your product to the person whose card you draw from that fishbowl.

> **You must chart every single response you get from every outlet you try. Otherwise you're asking for trouble.**

FOLLOW-UP, FOLLOW-UP, FOLLOW-UP

Probably the biggest mistake a novice makes is testing the advertising waters with a little ad here and a little flyer there and neglecting to keep accurate records of who he contacts and how many responses he receives. How in the world are you supposed to evaluate whether an ad has worked for you if you don't keep track of your responses? Estimates aren't good enough. You must chart every single response you get from every outlet you try. Otherwise you're asking for trouble.

A Cautionary Tale

An inventor attended a consumer fair one spring and sold some of his gardening products. The show coordinators asked all exhibitors for extra flyers to place by the door for attendees who didn't have an opportunity to stop by all the exhibits. This exhibitor put 250 flyers by the door.

In the fall of the same year, the inventor ran a small ad in a local newspaper. The first week or so didn't bring much response, but then sales picked up dramatically. After selling about 50 items, he decided to run a larger ad at four times the cost. He was extending his budget as far as he could, but since the smaller ad was so successful, he felt he couldn't lose.

When the larger ad ran, he got next to no orders and found himself out of money. What went wrong?

He didn't chart his responses. He assumed the calls he got in the fall were from the ad he had placed. But in fact, a woman had taken his flyers from the consumer fair in the spring, had tested his product, and had then included it in a booklet that listed top gardening products. She sold the booklet as a "helpful hints for gardeners" book. She included "reviews" on interesting or very helpful new products. She didn't sell anything. The booklet included flyers from the consumer fair so her readers could purchase any of the products they were interested in. Because he didn't chart his responses, the inventor didn't know anything about the booklet, and he ended up wasting money on advertising that did nothing for his sales.

Know who your customers are and how they found out about your product.

How to Track Your Customers

Don't get caught like this man did. Know who your customers are and how they found out about your product. If you take phone orders, ask, "How did you hear about the Scoop-O-Saurus?" If you are going to run more than one ad at a time, or more than one time in the same publication, you can place an identifying number (or word or phrase) inconspicuously in the ad. Then ask your caller, "What is the number in the bottom right corner of your ad?" If you do mail orders, you can include something like "Attention: Dept. 43" or "Attention: Scott" as part of the mailing address. Change it for each ad or flyer so you'll know which one your customer saw. If you are using flyers that have tear off and mail in order forms, you can simply print them in different colors as a designator.

As you garner responses to your advertising, you'll want to calculate exactly how much each response or order cost you. In order to do this, you must also keep accurate records on the expenses of each ad, flyer or commercial. If you use an 800 number, include those charges, as well as costs for printing, typesetting, photography, and postage in your calculations.

Keep in mind that some consumers who see your ad may not purchase the product right away. In fact, it's been estimated that the average person needs to be exposed to a product three times before making a decision to purchase it.

This doesn't mean you won't have sales the first time you run an ad; you surely will. But the majority of your sales will come after repeated exposure to the product, which means you must pay for that exposure over and over again. So when you calculate your costs and response rate to your ad, remember that you are also building customer awareness, name recognition, and credibility with your advertisements.

CHAPTER NINE

HOLD THE PRESSES

Press Releases and the Media: Print, TV, and Radio

Advertising, while often expensive, is also often very effective. It helps to presell your customer and create a demand for the product. If retailers receive phone calls from people asking whether they carry your product, you're going to have an easy time placing your product in their stores. Wouldn't it be great if there were some way to gain all those advantages without going to the expense? I'm pleased to tell you there is. It's called *public relations*, or *PR*. Public relations is the use of the media to let people know about a product, event, person, etc., that is free to the person submitting the material. (When the president takes a vacation, we find out about it through public relations, not because the White House paid for an advertisement.)

THE MAGIC OF PR

When you consider that a full-page color ad in a major publication can run anywhere from $20,000 to $75,000, you begin to look for alternatives. Sometimes alternatives will even come knocking at your door.

When stores mentioned in a newspaper article ran out of my product within 24 hours, I learned the power of the media.

My first taste of publicity came pretty early in my business life. Someone from the *Los Angeles Times* called me one day and said he had seen me at a baby fair and thought my product and I looked interesting; would I be willing to be interviewed for a story? I was not only willing but thrilled.

The paper sent out a photographer, who took pictures of my daughter with her Little Shirt Anchor on. The photo published with the story showed a baby's bottom, but that was *my* baby's bottom (a shot only a mother could be proud of).

Then the reporter called, and we did an interview over the phone. A few days before the story was to come out, the reporter called me back to say he needed the name of a local store that sold the Little Shirt Anchor. I gave him the name of a store with a couple of locations, one at each end of town. Maximum convenience for customers meant maximum sales for me.

When the article came out and my family and friends saw it, I was a star for a day. But the real payoff came the next day: Both stores mentioned in the article sold out of Little Shirt Anchors within 24 hours, and they called me to beg for more as soon as possible. That was when I learned the power of the media.

(text continued on page 166)

A Lesson Learned: Take Advantage of the Power of the Media to Sell You Product

The power of the media came to my aid when I wanted to get into a chain of specialty stores in Arizona called Irwin's For Children. I had sent the owner of these ten stores sample after sample of my Little Shirt Anchor, but she failed to be impressed.

I sent her some for her daughter to use, but her daughter was not impressed. This was contrary to all of the response, from my stores and customers, so I decided she needed some incentive to put the product in the store.

I called the Arizona Republic *newspaper, which is distributed all over Arizona. I told them I used to live in Phoenix, which was true. Since moving out of the state I had invented a product that I now sell mail order to a lot of Arizona residents. Wouldn't that be a great story? As it happened (timing is everything), a reporter was working on a story about local inventors. He had two so far, and he would be happy to include me. Could I send a picture right away? Why, yes, I could. Could I give him the names of some of my satisfied customers in Arizona so that he may get some quotes? Why, of course.*

As we came down to the eleventh hour, I told him his readers might want, as a result of reading his article, to go out and purchase a Little Shirt Anchor right away. "I've got an idea," I said innocently. I told him about this chain of ten specialty stores all over the state that carried the product. Most moms would already be familiar with the stores, so if he included the store name in his article, his readers would not bother him by calling the paper, trying to find Little Shirt Anchors. He agreed.

Now all I had to do was to tell Mrs. Irwin. The article was to come out in Sunday's paper. On the Monday before I called her. I told her that the paper was going to print her store name as a place where people could go to purchase a Little Shirt Anchor. I was sorry to do this to her, but the story would not run unless I included a store location. Could I please send a couple of dozen to each store, at my expense, just to have in case someone did want to buy one? I also said she could return any unsold merchandise, and I would bill her for only the Anchors that she sold.

(continued on next page)

Stores like to see their name in print, so she was happy, though she was sure none would sell. The story ran. On Sunday night, four of the stores called me directly with an urgent message: "We are all out of stock, please send more ASAP." One store called me to say they had never heard of me or my product, what was all this about? Obviously they had yet to get the Little Shirt Anchors out of the box in their back room. Irwin's For Children became one of my best accounts.

If you're going to be written up in a newspaper, use the opportunity to try to get your product into some local stores by telling them you'll put their name and store location in the article. (Check with the paper before making this promise, however.)

The secret to getting publicity is making your product—and yourself—newsworthy.

One Part Luck, Nine Parts Hard Work

Most PR doesn't just fall into your lap; it's a lot of hard work. But it can pay off in a big way. And one of the best things about it is that the more you do, the more exposure you get, with less and less effort. After I appeared with my catalog and product on "Good Morning, America" (more on this later), I received calls from several magazines and newspapers, other television shows, and even radio stations. Quite often the media will use other media sources for their editorial sections. And it's all free! Well, almost free. You do incur the cost of printing your press releases, buying postage to send them out, and making phone calls to follow up. But it's a far cry from what you'd pay to run even one advertisement.

That glossy magazine you could never afford to advertise in may be very willing to "advertise" your product for you in the form of an article or a new product review. A TV or radio show may be happy to expose your product to hordes of listeners or viewers. The secret is making your product—and yourself—newsworthy.

Sometimes publicity will fall into your lap, but hard work and persistence are what really make it happen. If you want real coverage, you must be aggressive.

THE PRESS RELEASE

Before you ever get any free publicity, you have to make people aware of your product. The way to do that is by sending out a *press release* or *press kit.*

A press release is a short article about you and your product that you mail to the media in hopes of sparking a story. There's a right way and a wrong way to write a press release. The wrong way begins something like this: "My name is Judy Ryder. I am a mother who invented this great new widget for using in the garden."

The first thing to remember is that all media have one thing in common: They are all designed to entertain.

What follows, generally, in a poor press release like this is a lengthy explanation of the product, followed by irrelevant information ranging from how long you have been an inventor to "My Uncle Harry gave me all the money I needed to get started." If your Uncle Harry is Harry Belafonte, Harry Hamlin, or Dirty Harry, that's interesting. If not, you are filling your press release with a lot of empty, useless facts.

The number one thing to remember is that all media—TV, radio, newspapers, magazines—have one thing in common: They are all designed to entertain. So first and foremost, your press release must tell them why you and your product are entertaining.

Don't think that a great new product is all you need. You can buy advertising space for your great new product, but to get editorial space, you need a story behind the product.

Finding Your Story

The first thing you must do is decide what makes your product—and you—worth writing about.

I discovered very early on that while my Little Shirt Anchor was interesting enough to make many of the "What's New" sections of publications, what really sparked editors' interest was finding out that a parent had invented it. That opened the door to a major story about parent-invented products. And it gave me a PR edge over my corporate-giant competitors. Playskool doesn't rate a story on how it got its latest product to market—who cares? But when a parent comes up with an idea and gets it all the way to the marketplace, that's a story.

Is there a story in you? Maybe you're an only child and you invented a blow-up sibling. Okay, a little ridiculous, but you get the idea. How did you come up with your idea? What motivated you to think about the problem and come up with a solution?

You need to come up with story angles that make you and your product newsworthy.

When you write a press release, you will need to come up with story angles that make you and your product newsworthy. The more ready-made stories you can feed to editors or producers, the more likely they are to give you the publicity you want.

Angling for Coverage

In his press release, an inventor who created a bumper, or cushion, that goes around the edge of the bricks on fireplace hearths to protect kids from being hurt came up with some ready-made story ideas to make it easy for the media to write up an article on him and his invention. In the press release, the inventor planted the seed for several story ideas:

- **Ways to babyproof a home:** latches on cabinets, protective coverings for outlets, and a cushion for sharp fireplace bricks.
- **A seasonal story idea on how to get your fireplace ready for use after the winter:** how to select a chimney sweep, products that clean the soot and stains off the brick, and a childproof bumper for the time of year when activity around the fireplace increases.

• **Safety around the fireplace:** avoid accidents by using self-starting logs, safety grates, spark resistors, and a cushion for the hard brick.

You might be wondering why you would want to mention other products or services in your press release. Why share the limelight? I'll tell you why: You will get far more press this way. Rather than getting a tiny blurb, or, more likely, no press at all, your product can be part of a larger feature that will be read or seen by many more prospective customers. If your product gets two paragraphs out of a four-column article, or 30 seconds out of a four-minute segment on a TV show, isn't it better than not being mentioned at all?

In your press release, plant the seed for several story ideas. Make it easy for the media to write up you and your invention.

A Sample Press Release

The presentation of a press release is very important. It means the difference between someone reading it and someone throwing it out without a second glance. Remember, the person you send your press release to is probably receiving hundreds of these every day; you must make yours "reader friendly."

I've included some press releases in the Library of Resources to give you some ideas about layout and content.

What to Include

You'll notice that at the top of the press releases in the Library of Resources is *the date* the material was prepared; this is very important because many times editors and producers will file information for use later on, and they won't use it if it's undated or out of date.

It also states *the contact person's name and telephone number.* Again, very critical. You don't want someone to have to track you down if they want to use your product or information. (In fact, they won't track you down; they'll toss it.)

Next comes a *summary*: a *brief* sentence or two summarizing the product and the main story. This should contain only the main facts, yet be descriptive enough to spark your reader's interest.

Although they are not in the sample press releases, you may want to include *headings, or titles,* for story ideas you've thought up. Again, aim for something attention-grabbing.

You'll also want to include a straight *fact sheet* that doesn't have any hooks, listing the product name, use, colors, sizes, suggested retail price, etc.

Finally, list any other items you've included, such as *photos, press clippings, or samples*. Do not send samples unless specifically requested to do so. You'll just waste a lot of money on postage and product (unless it's a very inexpensive item).

Put each little story idea on a separate piece of paper, with its own heading, in the order it is listed on your front page.

A good layout will make it a breeze for readers to find the information they're looking for.

Grab Their Attention Fast

People looking at the first page of your press release should know exactly what the entire package includes within 15 seconds. They might pull out the product information sheet and give it to the new product reviewer. They might notice the article idea on sprucing up the fireplace and give it to another editor who's working on a story about getting your home ready for winter. And because you gave them a package that is so complete, they don't even have to think about it! You started their work for them, and they will recognize that you made their job a little easier.

There's an oft-heard rule about keeping a press release to one or two pages. It makes sense; no one will read 15 pages of anything! But with a good layout like the one above, you *could* include 15 pages and still make it a breeze for readers to determine what they're interested in and where to find it in the release.

Getting Your Release Read: Magazines and Newspapers

Once you've written your press release, you have to get it into the hands of the right people. Are you angling for magazine coverage? Scour the newsstands for any publication that deals with your story ideas or products similar to yours. Buy a copy or copy down the publication's address and phone number (usually at the bottom of the masthead—the section that lists the publication staff).

If you want to focus on newspapers, libraries are your best resource. Ask at the reference desk for a publication that lists national newspapers, which will include addresses, phone numbers, and circulation information. You'll also want to seek out every newspaper in your local area.

When you call publications, indicate that you'd like to send in new product information and be sure to get the name and title of the appropriate person to send it to, and confirm the address. Check the spelling of the name, even if it's a common one. If there's any doubt, ask whether the person (Chris, Pat, Tony, . . .) is a man or woman. These small details can make a big difference in how your information is received.

Once you've written your release, you have to get it into the right hands.

Mail your press release in a large envelope, big enough to hold the information without having to fold it. Remember, appearance and neatness count.

Radio and Television

Researching PR possibilities in radio and TV is a little different, but your press release itself and your follow-up techniques will be the same.

Call the local TV or radio station that airs the show you want to be on, and ask for the name of the producer you should send ideas to. Get the same sort of detailed information I outlined above in

Getting Your Press release read. *Working Press of the Nation T.V.* and *Radio Directory* by Reed L. Sevier lists TV and radio stations nationwide.

A variety of story ideas will really come in handy when you send your release to TV shows. Unlike print publications, which can slip in a quick product release, TV shows do their programming in "segments," which are of a more substantial length. You need to make sure you have plenty of feature-story ideas for them.

Here are some sample story ideas that I have used. They all revolve around a central theme, each with its own character and style:

• *Mothers of Invention:* Average moms who saw a need and invented a solution.

• *Fathers of Invention:* Dads who took parenting one step further to develop a product to make life with a youngster easier.

• *Mother Invents Baby Product, then Launches Catalog of Parent-Invented Items:* One mom's story of taking her own invention and parlaying that into a catalog of the most useful "parent-invented" products on the market.

• *The Mother of Invention:* This mom started with one invention, and now makes a living inventing and helping other inventors to realize their dreams.

• *Parent-Invented Products:* Unique and exciting new baby products on the market, all invented by parents themselves.

• *Women Entrepreneurs:* Six successful women who have all started businesses at home.

Radio shows are much less likely to be interested in segment ideas and more likely to want to put some of the focus on you, via a telephone interview or in-studio interview. (More on handling this later.)

TV shows may be more interested in your product than in you. They may ask you if you know of any other new products that can be grouped together for a segment. Be prepared with as much information as you can get your hands on; make it your business to know the answers to any questions they may ask. And don't ever tell anyone "no." If they want you to locate a polar bear to demonstrate your new ice cube trays, find one!

Promptness Makes a Difference

Whether you're working with print or broadcast media, it's crucial to be conscientious about deadlines. If someone requests samples or photographs or anything else from you by next Tuesday, make sure it gets there Monday. Don't be late, even if you have to drive there yourself to deliver it.

Record Keeping and Follow-up

Be sure to make a list of everyone you send a release to and when. Then wait for the phone calls to come in. By the way, if you need to be away from your phone for extended periods during business hours, be sure you've got an answering machine or use an answering service; it's essential for a businessperson.

Whatever you do, don't get discouraged. If they can't use your information now, they may want it in a few months.

The waiting period should be about two weeks. Then you can start making follow-up phone calls. Don't put pressure on anyone; just find out whether your press release was received and whether any further information, such as photographs or samples, would be helpful. If they say no, tell them you'll follow up again in several weeks. And do it. Be politely persistent.

After your initial follow-up call, you'll want to wait about six to eight weeks before contacting anyone again. Be easygoing; say you're just calling to see if you can be of any help. If you have any new information about yourself or your product that wasn't included in the press release—if, say, your product wins an award or you get a new, noteworthy account—mention it briefly, then follow up with a letter. If you've gotten other press coverage, mention that too, and send them a copy of the article or broadcast announcement.

Patience Pays Off

Whatever you do, don't get discouraged. Even if those you contact can't use your information right away, they may well be interested in two or three months. Your follow-up calls will keep them aware that you and your product are out there. Of course, if publications tell you, "Don't bother to call us back—we'll call you," you may want to lay off for a while; but you can still send them news of other press or new accounts.

Before I finally got on "Good Morning, America" with my Little Shirt Anchor, I called the same person at the show every six weeks for 18 months! One day, instead of hearing "No thanks, not right now," I heard, "Would you be available a week from Wednesday?" It made all the calls worthwhile.

It wasn't just persistence, though, that got me on "Good Morning, America;" it was being in the right place at the right time. When I called, I told the producer I was going to be in the New York area on another program and just wanted to let her know I would be available—that might have made the difference.

Remember, you always have the right to decline to answer a question.

PERSONAL AND TELEPHONE INTERVIEWS

Telephone or personal interviews for publications are far less stressful than TV interviews, of course. Nevertheless, they require equal amounts of preparation and professionalism.

You can dress comfortably for a non-TV interview, but you should be prepared to have your picture taken. On more than one occasion, newspaper writers have come to my home with an unexpected photographer in tow. You won't always get advance warning. So make sure you and your environment look presentable.

Whatever questions you answer should, be answered truthfully; but remember that you always have the right to decline to answer any question. If you're uncomfortable talking about exactly how many units of your product have been sold or exactly how much you spent on your packaging or where you got your money, simply express your discomfort to the interviewer. Ask if the topic can be dealt with in another way or if it can be left out of the article altogether. (The same advice applies to broadcast interviews: You will usually be given a list of questions ahead of time, and you can voice any concerns before the broadcast.)

TV INTERVIEWS

A TV appearance can bring the shivers to many knees, mine included. I did several shows, including "Hour Magazine," "Success Stories," and "American Entrepreneur," before I went on "Good Morning, America." But those were all taped shows. Sure, they were in front of a live audience, but I felt secure in the knowledge that if I sneezed in the host's face, all of America would not see me.

During an interview, follow the host's lead—and be ready or anything.

My "Good Morning, America" appearance, however, was live. Whatever I said and did was what a national audience would see. Yikes! It was very exciting, but if I hadn't had a good friend with me for moral support, I would have crawled out of my skin. I must also say that Joan Lunden was wonderfully friendly and really helped ease my fears in those nerve-racking seconds before we went on the air. After that show, I felt like I could do anything. I did a couple of other live appearances including the "Home" show and felt much more comfortable.

Small-Screen Preliminaries

Before you go on a TV show, you will be interviewed, usually over the phone. This preliminary interview will be the basis of the host's scripted interview.

Often you will also receive a script and will be asked to study it. But generally, whatever you said in the phone interview is what they want you to say on air. If you told your phone interviewer you came up with a new type of elephant prod because you used to work in a circus and got stepped on by an elephant, don't go on the show and say that many different incidents led to your invention. They want to hear about the circus!

That said, don't expect the host to follow the script too closely. It's just meant to be a guideline. More often than not, your comments will pique the host's curiosity and lead him or her in a direction that was not covered in the script or by the preliminary interviewer. Be ready for anything: Follow the host's lead and don't anticipate the next question.

Dos and Don'ts

Do:

- Be concise and to the point when answering questions. It's amazing how much you can say in 10 seconds of carefully chosen words—and how little you can say in 30 seconds of nonstop prattling.
- Listen carefully to your producer for cues. If you will be handling props, the producer may instruct you to do a product demonstration first, then discuss yourself and the product afterward. Your producer may instruct you to stand or move around the set on a given cue, or perhaps give you a line or two on a teleprompter to read as an opening to the segment.
- Listen carefully to the host so that your answers correspond to the questions. Remember that old caveat: Be yourself.
- Smile, smile, smile, and smile. Nothing will do more to hide your nervousness or discomfort than a pleasant smile on your face when you're introduced and while you're being interviewed. Obviously, if you are talking about a serious subject, a huge grin may be out of place. Otherwise it can only help make you look and feel good.

Don't:

- Interrupt or correct the host. Remember whose show it is. It's not yours.

• Slam another product on the market or mention brand names or store names (unless you have permission from both the show and the store).
• Mention another TV show or network.
• Look at the camera or monitors unless you are told to do so.

Smile Mileage

When I did a show on the "Financial News Network," I was sent to wait with the other guests in the "green room." I was the only one who had been on a TV show before, and I felt very comfortable. One of the other guests, as it turned out, was the publisher of a magazine I'd been trying to get to publicize my product. He seemed very nervous, so I did my best to make him feel at ease and reminded him to smile. I told him it would make him look comfortable even if he wasn't.

It's amazing how much you can say in 10 seconds—and how little you can say in 30.

As we waited for our turn, we watched the show on a monitor. Another fellow who had been sitting with us earlier was being introduced by the host. He looked dazed and terrified—and barely aware that the camera was on him as the host was speaking. I turned to the publisher and said, "See what I mean about the smile?"

When the publisher went on and was introduced, he had a nice smile on his face. Afterwards, everyone told him how well he had done. The next edition of his magazine had a feature story on me and my product—another case of being in the right place at the right time!

The Show Must Go On

Doing live TV means expecting the unexpected and being prepared for things to go wrong, especially when you use props. Here's an example of the sorts of fun you can expect:

I appeared on the "Home" show to publicize the Little Shirt Anchor, my Snap Laces, and some other new products sold through my company, including one called the Soothe 'n Snooze Bassinet, a motorized bassinet that has a patented movement to help put babies to sleep. We had done what's called a *pre-tape*—footage filmed before the show—demonstrating the products being used with small children, but I was also going to give a live demonstration.

When I began expounding on the Soothe 'n Snooze's patented motion, I lifted the bedding to expose the keypad to camera number two for a quick closeup of the controls. I pressed the "on" button—and nothing happened. I pushed it again, and then again—still nothing. As this happened my monologue went something like this, "Let me demonstrate the patented motion by using this simple keypad. You just press the button, one quick step, and—apparently it's not plugged in!" Then I went on automatic pilot, gesturing to show how the bassinet *should* be moving, hoping for a miracle, while Sarah Purcell commented on the pitfalls of live television with a smile on her face.

Whatever happens, don't waste your moment in the spotlight.

Just in the nick of time, the pre-tape came on to save me. The segment turned out great, and I got some terrific exposure for my products. The moral? Whatever happens, don't waste your moment in the spotlight. Don't waste time trying to fix a problem; work around it. Remember, you only get one shot in such situations.

What to Wear

What should you wear for a TV appearance? It may seem like a small detail now, but you'll change your tune the night before or the morning of.

There are some general rules: Don't wear all white, all black, loud prints, or complicated patterns. Do wear something that you are very comfortable in. Women who plan to wear a dress should bring an extra pair of pantyhose. To avoid wrinkling, bring your TV clothes to the studio and change there.

Usually the studio will do your makeup and hair. But always check; once I showed up to do a show after a bout with the flu, expecting to be turned into a reasonable-looking human being, my face devoid of color and my hair uncombed. Alas—the studio did not have hair and makeup people!

RADIO INTERVIEWS

For radio shows, just as for television, you'll be pre-interviewed before the real thing. However, I've found that radio hosts are far more likely to stick to the scripted questions than are TV hosts—probably because they can keep the script in front of them during broadcasting.

If your radio interview will be conducted over the phone, it's particularly important to be well prepared, clear, and concise. And make sure you won't have any interruptions during the call.

MAKING THE MOST OF YOUR PR

Every time one of your press releases makes it into a newspaper or magazine story or turns into a TV segment or radio spot, make sure all your customers, your reps or your distributor, and stores that carry your product see it or hear about it. If you're going to be written up in a newspaper, use the opportunity to try to get your product into some local stores by telling them you'll put their name and store location in the article.

If you're working on a big account, let them know of any publicity you get as well, so they'll be aware of how hard you work to get your product in front of people.

Be Prepared for the Customer Onslaught

When you get a newspaper or magazine write-up, ask if the story can include a phone number so customers can contact you. For a TV appearance or radio show, make sure the station has your phone number at the switchboard. If your product is in a mail-order catalog, you may want to use that number so customers can order easily.

Make sure you are equipped to handle the orders and inquiries that will come from your appearances and write-ups. One paragraph in a

---

A Lesson Learned: Don't Neglect to Keep Your Support Informed

At the time I was invited to appear on "Good Morning, America," my husband and I had just begun to gear up for our mail-order catalog business. He had written a record-keeping/invoicing program on our home computer, and he was answering phones and filling orders. Three business phone lines rang into our house, all on rotation. That means if one line was busy and a call came through, it would automatically rotate to an empty line.

In the middle of all this, I left him to go to the Big Apple to appear on national TV. As I sat backstage at "Good Morning, America" having makeup applied, my husband was sound asleep in our California home. (The show is broadcast live on the East Coast and taped to be shown later in other time zones.) When I appeared on the show, it was 5 a.m. his time.

My husband had asked me to call him before I went on, but I was so nervous that I forgot to call him. The minute the interview ended, I dashed to a phone and dialed home. But the lines were already busy. That's how fast the audience reacted to the segment.

Well, however good that busy line made me feel about the sales coming in, it also made me realize I'd let my husband down a bit. He didn't quite share my excitement, at least not immediately, because he was sound asleep when all three phones started ringing. I had not called to warn him I was going on the air right then, and he didn't know what hit him.

---

major magazine might keep you on the phone nonstop for four weeks. One appearance on a major TV show and you'll be swamped for days. Be ready with either information to send or products to sell. Have on hand a list of the stores that carry your product so that you can give store names and addresses to callers asking where they can buy your product. If you're publicizing a mail-order product, be sure to let the catalog know in advance so it can stock up on your product.

When those orders come flooding in, all your PR efforts will be rewarded.

CHAPTER TEN

BUSINESS BASICS

Sales Transactions and Bookkeeping

One of the first things you should do when starting up your business is to develop a record-keeping system that's easy to maintain—and then start maintaining it. This chapter will give you an overview of the basics, a quickie course in the ABCs of sales and bookkeeping from someone who absolutely hates the thought of balancing her own checkbook.

ORGANIZING YOUR FINANCES: ACCOUNTS RECEIVABLE AND ACCOUNTS PAYABLE

Any business that involves selling will be concerned with two types of monetary transactions: paying for goods or services necessary to operate your business—*accounts payable*; and receiving money for goods or services supplied—*accounts receivable.*

Ideally, your payables will always be less than your receivables. But especially when you are first starting out, that may not be the case. Calculating these two figures every month will tell you where you stand financially.

SALES TRANSACTIONS

Here's a brief rundown of how sales are transacted:

If a store wants to buy 50 units of your product, it must let you know ahead of time by giving you a *purchase order,* or *P.O.*. Some stores will send you a written purchase order. These will always have a number on them that helps stores keep track of what they ordered and when.

If a store calls you and place an order over the phone, you may be given a P.O. number verbally, or you may not get one at all. Many companies will not deliver any products unless they have a purchase order from the customer, but usually this applies only to products that must be specially handled or that are custom made.

P.O. Info

A purchase order may contain some or all of the following information:

- store name and address
- shipping address
- the buyer's name
- the purchase order number
- the sales rep's name

- the date the order was placed or written
- a description of what is being ordered
- stock numbers or item numbers
- quantity
- price
- payment terms
- how the order is to be shipped
- when it is to be shipped (e.g., ship by November 11 or ASAP)
- date by which the order will cancel if the product has not been received
- any special instructions or comments

There's nothing worse than shipping an order and then having your account call and say, "I don't remember ordering this."

If you do take an order over the phone, you will need to ask most of these questions to get a complete order. There is nothing worse than shipping an order and then having the account call and say, "I don't remember ordering this." If you didn't mark the date, or ask the store buyer's name, you are out of luck. Get all the information when the order is placed.

You may also use purchase orders to buy your supplies, so you'll get to know them as a customer as well as a supplier. You can purchase fill-in-the-blank purchase orders at a stationery store. If you use these, keep one or two copies of the P.O. and send the other one to your supplier. Keep a record of P.O. numbers in a notebook or ledger, making sure you note the date and the supplier name.

If you have a computer, you might want to set up your purchase orders in your computer in a database, or purchase some simple software. *Quick Books* by Intuiet, *Peachtree Accounting* by Peachtree, and *MYOB* (*Mind Your Own Business*) by Best Ware are three current programs. There are dozens available; check with local software stores for others.

Filling the Order

After you receive a P.O., the next step is filling the order. Take note of when your account wants the merchandise to be sent, and try to make that date if possible. If you absolutely can't, alert the account as soon as possible, and see if you can work out some other arrangement. Never say you can ship a product by a certain date when you know that you can't just to make someone happy or hold on to an order.

Invoices, Packing Lists, and Mailing Labels

Before you ship your package off, you'll need to prepare an *invoice*—a bill that tells your account how much to pay and what they have received. It should contain much of the same information that was on the P.O., but will also include a total amount due, which is the price for the merchandise plus shipping costs, a payment due date, and payment terms. And it should indicate the shipping date, quantity shipped, and quantity back-ordered (more on this later). You can buy invoices at an office supply store; make sure you get the sort that includes a packing list. Some software like that mentioned above will generate invoices, labels, and packing lists for you.

Never say you can ship a product when you know that you can't.

Mail the top copy of the invoice to the billing address listed on the P.O. The other copies are for your files.

Once the invoice is completed, remove the packing list, fold it up, and attach it to the outside of the box in a clear plastic envelope made for this purpose and available at stationery supply stores. (A packing list is a carbon copy of the invoice, that includes all information *except* pricing, which is blocked out.) Whoever unpacks the box at its destination point will check the packing list against the contents in the box. If everything is present and accounted for, the customer will keep the packing list until his invoice arrives so he is sure he is only being charged for the merchandise he received.

One more step and your package is ready to go. Fill out a *mailing label* (also available at office supply stores) that includes a complete mailing address and complete return address. Double-check your P.O. for a shipping address, which may differ from the billing address. That's it.

Delivery

When you're ready to ship your product, refer to your P.O. and see if your account designated a particular type of shipping. UPS is pretty standard, but you may have a store in Alaska that wants its order to be sent parcel post. Your store is going to pay for the freight cost or shipping, so you need to follow its instructions.

UPS offers three basic types of delivery: overnight, second day, and ground, this last being the fastest way to get it there by truck. You can contact UPS directly to find out more about the services they offer and even get yourself set up as a shipper so the UPS truck stops by your place every day to pick up packages. Or you can ship from a UPS office or an independent shipping source that offers UPS service. Be aware that if you use a separate store location or shipping service that offers UPS shipping, you'll be charged more than the UPS rate. That's how they make their money. But you can only charge your customer for the actual shipping charges. Any service fees are your responsibility.

Back Orders

Say you have an account that orders 36 units of your product—12 green, 12 pink, and 12 blue—and you only have the green and the blue in stock. What to do?

Send the items you have in stock, and indicate on the invoice that the pink items have been back-ordered and will arrive at a later date. Accounts will sometimes cancel back orders, but generally they will allow you to ship the products later.

Keeping Your Paperwork from Running Amok

Bookkeeping is many inventors' last priority. They haphazardly spend money, forgetting to keep receipts. They buy an invoice book from the local stationery store, then neglect to use it. When they receive payment for an order, they deposit the check and make only a mental note that the invoice has been paid.

Don't let this be you. You are now a businessperson, and this is no way to run a business.

If you can't keep up with your bookkeeping, find someone who can handle it for you. Ignoring or neglecting some other aspects of your business may mean a missed sale or a missed learning opportunity. The ramifications of neglecting your bookkeeping are much more serious. You may not only lose income and customers—charging someone twice for the same order guarantees an unhappy customer—but you will also make yourself a likely target for the IRS. Remember, you are required by law to document your business's financial dealings. And if you get audited and think the IRS will cheerfully go through your bags of papers, receipts, and documents, you are mistaken.

You can ignore some parts of your business without doing too much damage, but ignoring your bookkeeping has serious ramifications.

Multimillion-dollar corporations can't afford to ignore their bookkeeping, and neither can you.

You don't need an accounting degree to keep your files in order. You just need to take the time to do it and have a system that is simple yet efficient. When I started out, I was fortunate to have a bookkeeper friend, Tina, who was happy to stop by and balance my checkbook or figure out where I had filed something. But you don't need a Tina to do it!

Paper Files

Staying organized is at the core of good bookkeeping. Start out with some file folders and someplace to put them. You can buy an inexpensive cardboard file holder, or invest in a filing cabinet. Label your files neatly and keep them in alphabetical order.

What to file? Just about anything related to your business, including:

- licenses and permits for starting up your new business, including receipts for fees paid
- catalogs or brochures from other companies, arranged by company name or type of product
- copies of all correspondence; even if you have a computer, print an extra copy and stick it in a file
- individual files for each of your accounts
- separate accounts payable and accounts receivable files
- PR contacts, advertising, etc.; you may want to make a file called "follow-up" to help you remember what calls you need to make next week

Staying organized is at the core of good bookkeeping.

Filing Invoices

When you generate an invoice, one copy goes to the account, one copy goes in the account file, and a third copy goes in a separate file for accounts receivable. Once an invoice is paid, make two copies of the check before depositing it in the bank. One copy gets attached to the invoice in the account file, with an indication of when the check was received. (Don't rely on the date on the check. Customers often write out all their checks on one day, then mail them as they get money in.)

The second copy of the check should be filed along with a copy of the deposit slip from the bank so you have an accurate record of every penny that has been deposited into your account. No monies should go into your account unless you have a copy for backup documentation.

You may also want to keep a summary sheet in each account file, noting the date of each invoice, the invoice number, the amount, and when it was paid. This will enable you to quickly check how often various accounts buy from you and how long they take to pay. You may begin to notice patterns—late payment or orders that come every six weeks like clockwork.

Invoices that were in your accounts receivable file can be refiled in a "paid invoices" file. This way you'll know how much money you took in during a particular month.

Recording Invoices

All your invoices should be recorded in one ledger book, in numeric order. Just list the invoice number, the date, the customer, and the amount. You can turn to your ledger whenever you need to check how many orders you shipped during a particular period. It will also come in handy if you need to look up an invoice by invoice number.

Log all payments in your ledger as well. You may want to put a check mark by the invoice number, circle the amount, or add a separate entry to indicate that the invoice has been paid. Just be consistent.

Accounts Payable Files

For your accounts payable, keep a file of bills as they come in. When they are paid, note the check number, the amount, and the date on the bill. If you make regular purchases from a number of companies, make separate files for each. Invoices from infrequently used companies can be filed under Miscellaneous A-M and N-Z.

Computer Files

Some people prefer to set up all their files on the computer. Many reasonably priced software packages on the market can accommodate the simple bookkeeping you need to get started. If you do use a computer for record keeping, be sure to make backup files. You should still keep a hard copy of most transactions as well, but you can skip much of the cross-referencing that you might otherwise have to do manually.

IS YOUR BUSINESS REALLY A BUSINESS?

This seemingly simplistic question is actually rather complex—and crucial when it comes to bookkeeping and taxes. According to the IRS, there's a fine line between a hobby and a business. For any operation to be considered a business for tax purposes, owners must show that they intend to make a profit.

Ask a certified public accountant, or CPA, if what you are doing qualifies as a business and therefore as a tax deduction. You can also contact the IRS directly and ask for literature on this subject. The IRS has people available to help with your questions all year long, not just at tax time. State laws may vary, so be sure to contact your local government offices as well as the Franchise Tax Board and/or the State Board of Equalization, if applicable. If you obtain a business license or seller's permit, you can ask your local agencies for advice and assistance.

According to the IRS, there's a fine line between a hobby and a business.

If you qualify as a business, many of the expenses you incur may be tax deductible, even if you plan to license your product to a company rather than manufacture and market it yourself.

HOW MUCH CAN YOU DEDUCT?

If you own your own business, using a D.B.A., or Doing Business As status, the federal government, and most states, will consider all the money you earn in your business as personal income. That's like getting paid without having taxes taken out. Taxes will not have been taken out at the time you received the income, but taxes will have to be paid at tax time if you earned more than you spent.

Business expenses or tax deductions are credited against your income to come up with a figure on which you will have to pay taxes. If, that is, the figure is a positive amount. New business owners very frequently spend more than they make. In that case, there are no earnings to be taxed.

D.B.A.

When you start a business and pick a name, "Johnson Products" for example, you must run a fictitious name statement in a newspaper in the county in which you have your business stating that Laura Johnson will be doing business as (D.B.A.) "Johnson Products." Call your local papers for pricing. This is usually required before opening a business checking account.

With few exceptions, if you do not have proof of an expense, you cannot deduct it. For example, if you pay someone $100 in cash to design and print your business cards, and they don't give you a receipt, the IRS will probably not accept this as a business expense. It is imperative that you not only collect but *keep* receipts.

It is imperative that you not only collect but *keep* receipts.

Possible Tax Deductions

Business cards are a pretty obvious business expense, but there are many other not-so-obvious costs you can deduct. Some or all of the following expenses may be allowable as tax deductions, so long as they are directly related to your costs of doing business:

- mileage on your car (keep a mileage log in your car)
- dry cleaning
- professional memberships
- subscriptions
- telephone bills (portion)
- utility bills (portion)
- mortgage or rent payments
- tuition for business-related courses
- prototype costs (see chapter two)
- books and office supplies

- trade show and consumer show costs (whether you attend as an exhibitor or as a visitor)
- parking fees
- consulting fees
- professional services (attorneys, CPAs)
- meals, hotel, airfare, rental car, and other travel costs
- certain clothing
- computer software and equipment
- telephone answering machines
- gifts
- postage
- tools, equipment, and machinery

and more.

Keep track of *all* expenses related to your business, then ask your CPA or tax office if they are in any way deductible.

Credit and Collections

As a business owner, one of the most difficult things you need to do is install and follow a rigid program for credit and collections. What's difficult is the enforcement part.

The most difficult part of collections is learning not to bend the rules.

Maybe you'll decide to set a policy that no account with an overdue invoice will ever be shipped merchandise. Then one of your favorite customers calls—the one with whom you celebrated the birth of her first grandchild two weeks ago—and tells you her store needs an order right away and she'll put a check in the mail for the amount the store still owes you. You may be persuaded to send the order, then never get the check from the overdue invoice or the new one. Two months later your customer calls to let you know, with many tears and apologies, that she's going out of business. Suddenly you're left with two invoices you will never collect on.

This is just one example of why it's difficult, but often necessary, to say no. Sometimes, too, you'll want to bend the rules for the sake of business rather than sentiment: Can you imagine getting the biggest order your business has ever had, knowing that this store does so much business they are sure to follow with another order every few weeks, then having to refuse it because the store's credit is not satisfactory?

Shipping C.O.D.

When you receive an order from a new customer, you may know nothing about the account other than the business name and shipping address. But you need to find out more if that customer's going to get payment terms or credit with you.

Most companies make it their policy to send a new account's first order *C.O.D.—cash on delivery*—with payment terms to be established thereafter. This enables the manufacturer to check out the customer's credit, which takes some time, without holding up the shipment.

Shipping C.O.D. buys you time to check out a new account's credit without holding up the order.

To ship an order C.O.D., you simply need to alert your delivery company. If you use UPS, for example, it will pick up the package from you and deliver it to your customer, collecting a check or cash at the time of delivery. If the delivery person collects a check or money order, it will be forwarded directly to you through the mail; if the customer pays cash, UPS will send you a check. You may specify to UPS whether you will accept a check, a money order, or cash. Some companies will not accept a check without first checking their customer's credit.

Credit Applications

You will need to have your customer fill out a credit application. If you have purchased goods to manufacture your product, you're probably familiar with such an application because you most likely had to fill some out yourself.

You can type up a credit application yourself or purchase them in an office supply store.

In addition to obvious information such as store name, address, and telephone number, you also need to know the following:

• **the name of a contact person**, which will help you avoid getting the run-around if you need to call and follow up on an invoice.

• **whether the company is a corporation, a partnership, or a sole ownership.** If the company is a corporation, you want the name(s) and address(es) of the officer(s).

If the company is not a corporation, the store is nothing more than a person who has chosen to do business under a different name. That means the owner is really the one to whom you are extending credit. You have every right to know how to contact this person at a location other than the store; that's why you'll also need to know:

• **the owners' names, home addresses, and phone numbers.** Officers of a corporation are not usually personally responsible for debts the corporation incurs, but this information can be helpful if collection becomes necessary on a privately owned company.

• **credit references**, to help you learn about the account's payment history with other companies.

Don't be surprised if you get nothing but good reports. Companies don't usually list the names of people they haven't paid. Still you need to send a credit inquiry to each reference listed, asking if your customer has terms with them, i.e., whether they can receive merchandise and then pay for it in 30 days (net 30 terms).

Ask if the customer pays on time, and find out what the customer's *high credit* has been—the most it has owed the company at one time. This will help you determine how much credit you can extend.

• **the name, address, and phone number of the customer's bank, and its bank account numbers.** Although a bank cannot tell you how often your customer has bounced a check, you can call after receiving a check to find out if there are enough funds in the account to cover it.

• **the customer's DUNS number.** DUNS refers to Dun & Bradstreet, a company that keeps track of thousands of other companies, gathering information about sales, number of employees, size of facilities, etc., to determine a rating.

Contact Dun & Bradstreet and find out how to become a subscriber. This will give you access to its files on other companies. You may also want to get a Dun & Bradstreet rating for your own company; it helps when you're seeking your own credit.

One thing to keep in mind about Dun & Bradstreet: Most of its information about companies is obtained from the companies themselves. Dun & Bradstreet at this time does not double-check this information.

Credit information companies' fees may be high, but three or four uncollected invoices may convince you they're worth the cost.

Manufacturers Credit Information Companies

Another resource for checking credit is a manufacturers credit information company. Such an organization differs from Dun & Bradstreet in that the information is based on the report given by manufacturers who supply these stores. If a manufacturer has had a bad experience with a store, if they have bounced a check or not paid a bill, the manufacturer will have their own credit history for that store—this is what gets reported to a manufacturers credit information company. The stores have nothing to do with the information reported, which makes this a reliable and valuable resource.

Most of these companies are industry specific; in other words, if you're providing products for the beauty supply stores, a pet-store credit company won't help you investigate your clients. Contact the professional organizations in your industry for information about manufacturers credit companies. Their fees may be high, but three or four uncollected invoices will usually convince you their service is worth the price.

Remember, credit information organizations base their information on what subscribers report to them. If you decide to sign on with one, be sure to let the organization know if you have a credit experience with an account, so that other manufacturers can be alerted.

Evaluating Credit Information

After you get your credit information back from credit references, Dun & Bradstreet, and whatever other sources you use, how do you determine if terms should be extended and for how much? Unfortunately, I can't offer any hard-and-fast rules: This is an art, not a science. Common sense will guide you in the beginning; experience will chime in later, making such decisions much easier. In the meantime, simply review all your information and use your best judgment.

BAD DEBT

All businesses have to deal with unpaid, or uncollected, invoices, otherwise known as *bad debt*. But you can take steps to make sure your bad debt is as low as possible.

What if you get an order from someone with terrible credit? You can always ship C.O.D. Or you can request that the customer send you a check in advance, or pay half in advance and the balance in 15 days. Extend yourself only as far as you feel comfortable.

Collection Calls

When an invoice is 45 days overdue, you should send a letter or call that account. Don't back off. Persistence pays.

I know not everyone can stomach the idea of making collection calls. You may feel a little peculiar knowing that the very person

Collecting your payments is no less important than getting the sale in the first place.

who can assist you in becoming successful (the store owner who buys your product) is the person you now have to call and hound for a payment. But the reality is, making money is what you are in business for. And if a customer can't pay, you don't want that customer.

In the Library of Resources you will find some examples of collection letters that you may use or borrow from in creating your own.

If the invoice is *really* late—90 days or more—you may want to consider hiring a collection agency or attorney to collect it for you. Their fees can be as high as 50 percent of the invoice, so make sure you've done all you can to collect the payment first.

Just as with many other things in your business, credit and collections take time, planning, and effort. But no matter how busy you are with following up sales leads, PR leads, sourcing suppliers, checking competition, following trade shows, getting your printing done, and sending samples, you can't neglect your invoices. Collecting your payments is no less important than getting the sale in the first place.

CHAPTER ELEVEN

HOW NOT TO GET TAKEN

Avoiding a "Big Idea" Company

Some of the "help" offered to inventors is not help at all, but a way for other people to prosper on your hopes and dreams of success and financial independence.

THE COOKIE-CUTTER APPROACH

While I was struggling along on my own with the Little Shirt Anchor, I saw a few ads for companies that offer to help inventors make money with their idea. I never looked into them. I suppose I reasoned that I knew my product better than anyone else.

I also recognized fairly early on that what I needed to succeed with my product was specific knowledge, not a general, cookie-cutter approach: Inventors need experience and knowledge in the particular industry in which their product belongs. Every industry works differently. And there are so many types of product ideas, each one requiring a different strategy to make it successful. I couldn't understand how these companies could have one formula for making any product idea successful.

How can companies have one formula for making any product idea successful?

There Are No Guarantees

It is impossible to guarantee success of a product idea. No mortal being on the face of this earth can know if a product idea is going to be a winner. That includes the experts in the biggest companies around the world.

If success could be predicted, major corporations would never experience failure. But it's common knowledge that most industry giants—from car companies to baseball bat manufacturers—have all had tremendous flops.

HOW "BIG IDEA" COMPANIES OPERATE

My first direct contact with the workings of an idea marketing company—a "big idea" company, as I like to call it—was through a woman who came to me for consultation.

When she told me she needed my help but had already paid a company $10,000 to help her market her product, I couldn't believe my ears. Certainly none of my individual consulting clients ever paid that kind of money for my help. Some larger companies had, when doing detailed product evaluation, focus groups, and market testing. But never an individual.

My client—I'll call her Ellen—said she'd enlisted the services of a "big idea" company she'd found out about through a radio ad. After all the money she'd paid so far, she didn't feel she had gotten anywhere; although she had some time left on her contract with the company—and was still making payments to the finance company it had referred her to—she thought she should turn elsewhere for advice. She still believed in her idea

I asked her to describe to me in detail what the "big idea" company had done for her. I was very curious to see what sort of advice would be worth $10,000.

Drawing Inventors Into the Web

Ellen told me that after making an appointment with the company, she met with a vice president to discuss her invention. He—apparently—could barely contain his excitement as she described her idea to him. He was so enthusiastic, in fact, that he called in his higher-ups to come look at her product.

When a client told me she'd paid a company $10,000 to help her market her product, I couldn't believe my ears.

Two more men in suits came into the room; they, too, seemed extremely excited about this woman's idea. They estimated her first-year earnings at about a million dollars and said a product like hers could bring endless sales and unwaning popularity.

Even Skeptics Are Seduced

Now, try to put yourself in my client's place. *You* have always thought your idea had the potential to become hugely popular and successful. Suddenly, not one, but three executives from a company that looks at ideas all the time are telling you your expectations may have been too low. They're sure you're destined to go down in history because of this invention.

Maybe you're a little skeptical. You ask them how often they have told other inventors the same thing. They look around at each other and tell you—"in all honesty"—that they can't remember ever seeing a product that had this much potential.

Wouldn't you feel like you were sitting on top of the world? Like all your wildest fantasies were about to come true?

The execs take a break and offer you a beverage, maybe a snack. The VIP treatment. After the fluster has subsided a bit, they begin talking among themselves about all the potential markets for your product. They begin throwing big company names around and deciding who merits a glimpse of your product idea. Occasionally someone says, "But Bob, that may cost a little more." But by this point, you're thinking, who cares? If they're talking about the difference between making $1 million and $20 million, what's the big deal about spending a few extra thousand to do it right?

Imagine not one, but three executives telling you that your product is destined to go down in history!

The Heart of the Matter: Fees

Now it's time to talk about the company's fees. You've got such a great idea that they don't want to skimp on it; how much are you willing to pay for a ticket to the biggest lottery of your dreams?

The so-so success package is $2,500. All the execs agree that's not sufficient. Bob suggests the $5,000 package, noting that, after all, you're just a regular person, and you probably don't have the kind of funds to "do it right."

But Dave chimes in, "I can't believe our client wouldn't want us to use the safeguards to ensure her product's success when it's only a few thousand over and above that amount. Besides, we'll put her in touch with a finance company that will help her with the money—until the big money starts rolling in."

So before you know it, you're agreeing to pay your "big idea" company a fee of up to $10,000, and you're signing an agreement with a finance company to loan you that amount.

You walk out with a rundown of all the work this company is going to put into your product. You're so excited you are about to burst. With everything they're going to do for you—file your idea with the patent office and supply you with drawings; get your product into tons of trade shows; use their huge mailing lists to get your product in the right hands—what's a payment of a couple of hundred bucks a month?

The Great Swindle

After my new client, Ellen, had related her experience, I recognized all this hype for what it was: one of the greatest sales jobs ever.

Especially when Ellen described her product idea: a doll that looks like her, with a wardrobe of clothes with a big "E" on the chest (for Ellen). I couldn't believe it. It wasn't what I would call a unique idea, except for the fact that the doll looked like my client. But why would anybody want a doll that looked like my client?

What really made me mad was that this woman, an unemployed mother of four who barely had the money to feed her kids, was still making payments to the finance company that the "big idea" company set her up with. She was threatened with collection every time her payment was late.

All Talk, No Substance

What had the people at the "big idea" company done for my client in return for all that money?

They gave her a schedule that included a series of six-week goals that would carry her product along for the next two years—at which point her deal with the company would end. In six weeks the drawing would be done. Six weeks later her idea would be sent to the patent office. Six weeks later a list of target companies would be compiled . . . etc.

They had an artist draw her product, which she could have had done herself for $50. They had flyers made with the drawing on one side and a description of the product on the other. They did file her idea with the patent office, through a program called the Document Disclosure Program. It cost $10 at the time; you send them your idea, they stamp it "received," and they send you a copy back. The document indicates the specific date on which you filed your idea. It doesn't say anything about who may have thought of it before you.

They provided her with a list of the places to which they'd "submitted" her idea. Their version of submission, by the way, involved sticking a flyer in an envelope and mailing it to a company. The companies included a garden hose manufacturer, a tire company, and a board game company, among other laughably inappropriate targets. In short: Ellen got ripped off.

How much are you willing to pay for a ticket to the biggest lottery of your dreams?

The Upshot

I contacted the "big idea" company on Ellen's behalf and explained her financial situation. I was told very nicely that the company was performing all the work it had agreed to in the contract.

Ellen never again managed to speak to the three men who had raved about her idea, and she never got out of the deal.

FIGHTING BACK

Shortly after I spoke with Ellen, I was contacted by a couple who had also signed on with a "big idea" company. They were embarrassed even to admit they had done it—and had used their entire savings to do it.

I asked them how the company had handled them and what it had promised to do. The couple told me almost the same story I'd heard from Ellen. Unlike her, however, they had recognized pretty early on that they had made a mistake; now they just wanted to start over. They still felt confident that they had a good idea, and they wanted to make it work.

Unfortunately, although their idea—a child-safety restraint to be used in a car—had much more merit than Ellen's doll, it had one enormous flaw: It didn't comply with government regulations and therefore was not legal. The "big idea" company never told them that, of course.

Exposing the Enemy

I felt so sorry for these clients—plus countless others who had paid a lot of money for what they not only could have done themselves, but done better—that I decided to take action.

I asked my clients to provide me with copies of every document they had. Although these clients had gone to different "big idea" companies, the documents made it clear that both companies operated along the same lines.

I contacted the TV show "Inside Edition" and told them this scam needed to be exposed. They bit.

One of the show's reporters went to the company that had shanghaied the couple with the car restraint, armed with a really far-fetched, ridiculous product idea—and a hidden camera. The product idea was an automatic cruise control for a car, which locked down the gas pedal once drivers reached their desired speed. There is no way this product would pass the strict safety requirements the auto industry must adhere to.

The "big idea" company never once questioned or offered to investigate the product's legality.

But with that hidden camera, the "Inside Edition" reporter recorded the whole scenario, which played out just as my clients had described it. Sure enough, the men in the suits got all excited about this impossible idea, and never once questioned or offered to investigate the legality of such a product. They signed the bogus inventor right up!

Justice Is Served

As soon as the "big idea" company found out about the "Inside Edition" segment, it began calling my clients, pleading with them and promising just about anything to keep the segment from airing. It didn't work. In fact, an interview with the couple was included in the segment.

Since the segment aired, I understand changes have been made in these companies. But I continue to receive many idea submissions from them—envelopes with flyers inside, like those created for my clients. I throw them away without looking at them. And I'm not the only one. I have worked with research and development departments and executives in charge of product review for several major corporations, and I know the majority of them send these envelopes straight to the recycling bin.

PRECAUTIONARY MEASURES

I don't want you to get the wrong idea from all this: Some companies are legitimately in the business of helping inventors. I own one! And maybe even some of these "big idea" companies have helped an inventor or two get their product ideas seen and accepted by a company.

The preceding stories are meant not to make you paranoid but to remind you to be careful. When you consider hiring a company or an individual to help you launch your product, make sure you know *exactly* what services they will provide. Don't fall for generalities. If a company offers you a list of 60 trade shows at which it has previously exhibited products and at which your product could potentially be displayed, that's far from a guarantee that your product will end up at any of them.

Before entering into any agreement with a "big idea" company, take your contract to an attorney for review. You may pay the attorney $200 for this service, but that's a lot better than paying $10,000 for nothing.

Make sure, too, that you can get out of your arrangement if you are displeased with the company's services. If the company is confident that it can provide you with a service, it won't hesitate to agree to these conditions. If, on the other hand, a company tries to lock you into an expensive, long-term deal, you should definitely be suspicious.

Don't be blinded by talk of sure things or easy money. As I've said throughout this book, achieving success is not easy; it takes hard work and dedication. And sometimes you can only count on yourself for those.

REAL HELP

If you don't want to gamble on a "big idea" company, but you don't want to go it alone either, you can turn to a couple of other sources of support and advice: consultants and inventor groups.

If you don't want to go it alone, you can turn to a consultant or an inventor group for support.

Consultants

Consultants generally charge clients by the hour, for time spent talking to them on the phone or in person, and time spent working on their idea.

In my consulting business, I try to adapt my services to my clients' needs, and that includes their financial needs. In some cases I simply offer step-by-step advice to inventors on how to reach their goals, answering questions and supplying examples much as I do in this book. In other cases I'll do the work for them, which obviously costs more.

Since I can't guarantee the success of a product any more than the next person can, I never lock my clients into long-term agreements.

There are other consultants who think like I do, and if you do decide to use a consultant, look for someone who'll adapt their services to your real needs. Ask lots of questions and take the same types of precautions mentioned above before you sign on any dotted lines.

Inventor Organizations

There are many inventor organizations throughout the country. Check your local library. Also try *The Encyclopedia of Associations*—it has regional and state listing directories (use the main encyclopedia and look up "Inventors," get the names of associations and look them up in the regional and state directories). Your telephone directory has listings under "Associations."

Networking with other inventors can be very advantageous.

Inventor organizations can often refer you to useful resources. They may also offer newsletters or seminars and talks by professionals that will help you along your way. And networking with other inventors through such organizations can be very advantageous. You'll know you're not alone, for one thing, and you'll be able to learn from others' mistakes and good fortune.

LIBRARY OF RESOURCES

Business Plan

Sample business plans are easily gotten. There are entire books on them. They can be short and simple or very long and detailed. Here's part of a very simple one that my partner and I created for one of our new business ventures.

Primary Goals and Strategy

Our primary goal is to manufacture and distribute innovative items in the lucrative pet products industry.

In October 1993, we established distribution for one of our products (a child's "Handprint" version of "Paw Prints") in the juvenile products industry. We have also made presentations to buyers at national-level pet specialty chain stores for input on our opening product line of six products, which was well received and resulted in a test order of "Deco-food Containers" and "Pet Caddys."

Products

Our unique product line includes: "Paw Prints," a decorative tin with all the items you need to permanently cast your pet's paw print; "Pet Caddy," a hanging, netted bag to keep all your pet supplies in one convenient location; "Splash Guard," which helps to keep your bathroom dry while you bathe your pet; and large, decorative tins with original designs for storing pet food.

Examples of the products we plan to introduce later this year are: a line of pet birthday items, pool toys, safety harnesses, and nontoxic ant repellent strips. Our product line will offer both utilitarian products and lighthearted, fun products for real pet lovers.

Distribution

Our strategy for growth will include continuing to develop unique products, hiring sales reps nationwide to handle retail accounts, monitoring cost-efficient ways to handle products, and exploring alternative distribution methods, such as offering our Pet Caddy to R.V. supply and camping stores, and premium incentives with companies like PetCo for their "Birthday Club" promotion.

Projections	1994	1995	1996
Income			
Wholesale sales*	231,000	409,975	676,458
Freight income[1]	7,392	13,117	21,648
TOTAL INCOME	238,392	423,092	698,106
Cost of Goods Sold			
Returns and allowances	6,930	12,299	20,294
TOTAL SALES ADJUSTMENTS	6,930	12,299	20,294
Purchases—merchandise	138,600	225,486	338,229
Purchases—packaging	9,240	16,399	27,058
TOTAL COST OF GOODS SOLD	154,770	254,184	385,581
GROSS PROFIT	83,622	168,908	312,525
Expenses (Start-Up)			
Office supplies/artwork	2,700		
Office equipment	2,500		
Samples[2]	350		
Licenses/fees	800		
Product development[3]	1,100		
Office expenses	500		
Beginning inventory[4]	5,000		
Insurance	600		
TOTAL START-UP EXPENSES	13,500		
Expenses (Ongoing)			
Insurance	1,800	3,600	6,000
Commissions[5]	18,480	32,798	54,117
Freight	10,164	18,218	27,058
Office expenses	2,400	3,600	6,000
Salaries and wages**	10,400	20,800	42,640
Rent	2,400	4,200	6,000
Administrative	1,500	3,000	6,000
Utilities	900	1,500	2,400
Product development	1,2002,400	4,200	
Miscellaneous	1,500	3,000	6,000
TOTAL EXPENSES	50,744	93,116	160,415
Net Profit (Loss)	32,878	75,792	152,110

NOTE: The above figures do not include a salary for the owners.

* 1994 sales based on a penetration rate of 25% of the targeted customer base at a minimum order per month on four products. 1995 sales based on a 60% increase in customer base, while increasing the minimum order 15%. 1996 sales based on a 60% increase in customer base, while increasing the minimum order by 15%.

** 1994, 1 employee; 1995, 2 employees; 1996, 3 employees.

Notes for book readers:

[1] Freight income: Money that a company collects for freight, which is added to the company's customer invoices. These amounts are paid out by the company first, then re-billed on the invoices. It should not be shown as sales income or revenue.

[2] Samples: Products provided to reps or accounts to help generate sales.

[3] Product development: The costs associated with getting a product ready for sale, such as prototypes, professional fees, research fees, market and focus group testing.

[4] Beginning inventory: The cost of your first order of product.

[5] Commissions: Amount paid to sales reps, usually a portion of actual sales, as payment for their services.

We plan to concentrate on specialty stores the first year, which offer approximately 3,500 retail outlets nationwide, and mail order catalogs. In our second and third years we will concentrate on mass merchants such as Kmart and Target and warehouse and home supply chains such as Sam's Club and Home Depot. Both of us have experience dealing with accounts of this size in other industries.

Summary

Our diverse backgrounds and experiences in sales, marketing, public relations, accounting, administration, personnel, inventory control, purchasing, and product development provide us with the tools we need to be successful. We also have access to extremely low cost warehousing, supplies, labor, and home offices we plan to utilize as long as possible.

Patent

This is the patent granted me for the Little Shirt Anchor.

The United States of America

The Commissioner of Patents and Trademarks

Has received an application for a patent for a new and useful invention. The title and description of the invention are enclosed. The requirements of law have been complied with, and it has been determined that a patent on the invention shall be granted under the law.

Therefore, this

United States Patent

Grants to the person or persons having title to this patent the right to exclude others from making, using or selling the invention throughout the United States of America for the term of seventeen years from the date of this patent, subject to the payment of maintenance fees as provided by law.

Commissioner of Patents and Trademarks

Attest

United States Patent [19] [11] **Patent Number: 4,853,979**
Ryder [45] **Date of Patent: Aug. 8, 1989**

[54] **RELEASABLE SECURING MEANS FOR AN INFANT'S SHIRT**

[76] Inventor: **Judith A. Ryder,** 5064 Gaviota Ave., Encino, Calif. 91436

[21] Appl. No.: **197,706**

[22] Filed: **May 23, 1988**

[51] **Int. Cl.**⁴ **A41F 17/00**
[52] **U.S. Cl.** **2/326;** 2/332
[58] **Field of Search** 2/326, 327, 328, 332, 2/333, 334

[56] **References Cited**

U.S. PATENT DOCUMENTS

14,123	6/1896	l	12/333
311,111	1/1885	Eisler	2/326
755,188	3/1904	Tooker	2/326
863,970	8/1907	Ebersole	2/326
942,323	12/1909	Haywood	2/326
949,827	2/1910	Kurtz	2/332
1,320,641	11/1919	McCain	2/334
1,404,719	1/1922	Postl	2/326
1,638,304	8/1927	Guy	2/326
1,653,288	12/1927	Johnson	2/326
1,880,779	10/1932	Cahn	2/326
2,185,400	1/1940	Cohen	2/334
2,727,247	12/1955	Bailey	2/333

OTHER PUBLICATIONS

Maurice Gershman, "Self Adhering Nylon Tapes", *The Journal of the AMA,* p. 930, vol. 168, No. 7, dated Oct. 19, 1958.

Primary Examiner—Werner H. Schroeder
Assistant Examiner—Jeanette E. Chapman
Attorney, Agent, or Firm—George J. Netter

[57] **ABSTRACT**

A first elastic strip or ribbon is folded into V-shape and provided with two button-and-loop connectors at the ends for fastening to an infant's shirt front. A nonelastic strip is secured to the foldover point on the first elastic strip. A second elastic strip is formed into a V-shape and provided with button-and-loop connectors at its ends for fastening to the shirt back. Hook-and-loop connector halves are located on the nonelastic strip and base of the V on the second elastic strip for adjustable and releasable securement.

2 Claims, 1 Drawing Sheet

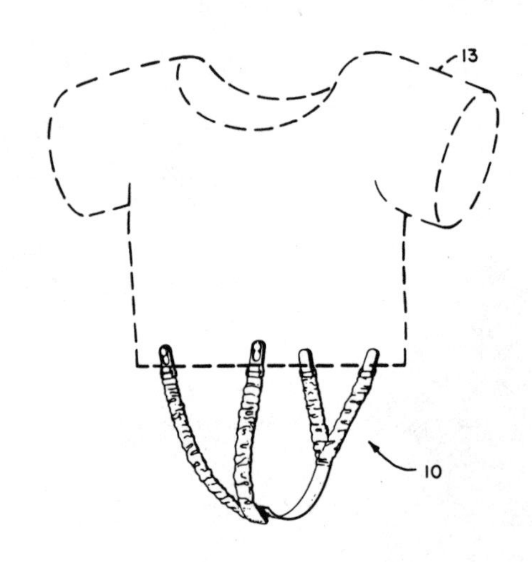

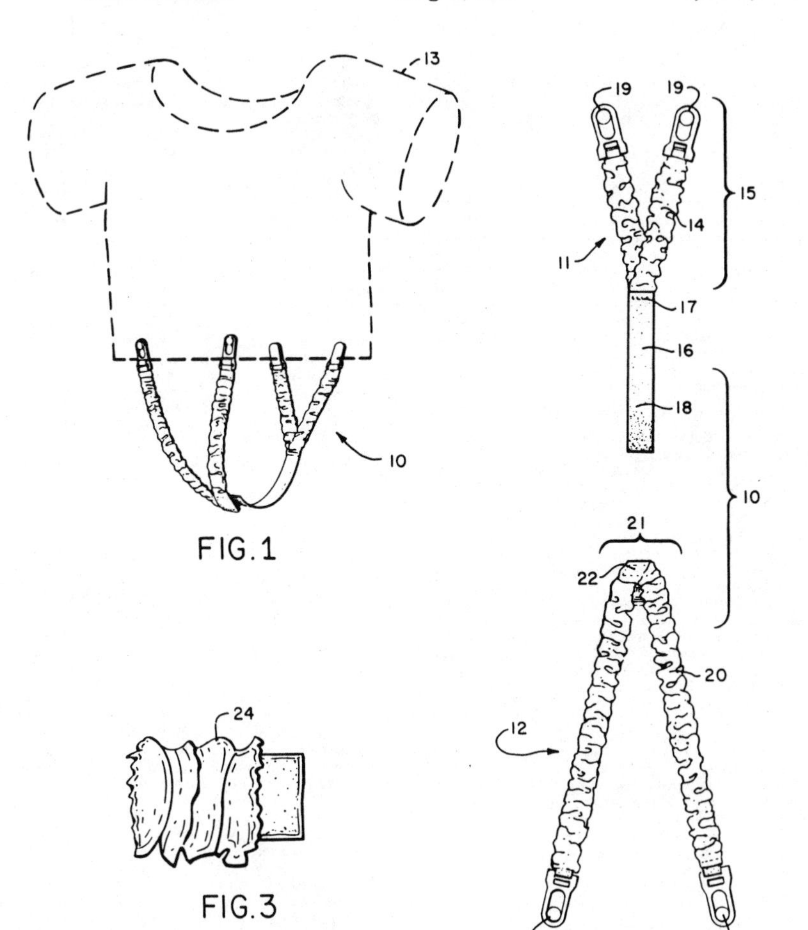

FIG. 1

FIG. 2

FIG. 3

4,853,979

1

RELEASABLE SECURING MEANS FOR AN INFANT'S SHIRT

The present invention relates generally to infant's clothing and, more particularly, to a releasable means for securing a baby's shirt in place while it is being worn.

BACKGROUND OF THE INVENTION

It is well known that an infant's clothing readily becomes twisted and pulled out of normal position while it is being worn, which is not only lacking in aesthetic appearance, but also is uncomfortable to the infant in as a result of knots and lumps being formed.

In the past, an infant's garments such as a shirt, for example, was typically pinned to its other garments or diapers so as to prevent the shirt from twisting, turning or riding up high on the infant, for example. However, the use of pins in this connection are not completely satisfactory in that they can open and possibly injure the infant. Also, a safety pin can, if applied repeatedly to the same place on the shirt, can end up tearing the garment.

SUMMARY OF THE PRESENT INVENTION

In the practice of the present invention there is provided means for securing an infant's shirt in place consisting of two separable parts. The first separable part includes first and second elastic strips on the outer ends of which are a releasable fastening means (e.g., button clips). The other ends of the two elastic strips are secured as by stitching, for example, to a single elongated strip of material on a surface of which is one half of a hook and loop releasable connector.

The second part of the invention consists of two elongated strips of elastic material which can be identical to that of the first part, one end of each being secured to the other so that forming two equal length arms extending away from a common interconnection in a generally V-shape. The outer ends of each of the elastic strips includes a releasable fastening means. At the crossover interconnection between these two elastic strips there is provided the complementary or mating hook and loop connector.

In use, the first part has its two releasable fastening means snapped onto the back of the infant's shirt at spaced apart points with the strips carrying the hook and loop connector half being located between the infant's legs. The second part of the invention is similarly releasably connected to the front part of the infant's shirt at a spaced apart distance greater than that on the rear and then the two hook and loop complementary connectors are suitably adjusted for the size of the infant and interconnected by pressing together.

DESCRIPTION OF THE DRAWING

FIG. **1** is a perspective view of the invention shown holding an infant shirt securely in position so that it will not ride up or otherwise become bunched.

FIG. **2** is a view of the invention with the parts extended and separated.

FIG. **3** is an enlarged, fragmentary, sectional view of a strip used in the invention.

DESCRIPTION OF A PREFERRED EMBODIMENT

With reference now to the drawing, the securing means of the invention, enumerated generally as **10**, includes first and second parts **11** and **12** which in a way to be described releasably and adjustably interconnects with an infant's shirt **13**. The securing means **10** serves

2

to hold an infant's shirt in normal wearing relation on the child and is contemplated solely for use where the child is also wearing a diaper.

The first part **11** includes an enlongated strip **14** (e.g., 5 inches) of elastic material folded onto itself to form a V-shape elastic member **15** with equal length arms. A strip **16** of a non-elastic fabric material is preferably affixed to the foldover point of the member **15** by a line of stitching **17**. One surface of the strip **16** is substantially covered with one-half of a hook and loop connector **18** which is sold commercially under the registered trademark VELCRO.

At the outer end of each V-shape member arm there is secured a button and connector **19** for making nondestructive connection with the shirt fabric.

The second part **12** of the invention includes a strip **20** of elastic material (e.g., **12** inches) which is folded onto itself to form a V-shaped construction **21** with two equal length arms. At the strip crossover point a second and complementary part **22** of the hook-and-loop connector is secured, preferably by stitching. Each end of the strip **20** includes a button and loop connector **23** which can be identical to the connector **19**.

Although the strips **14** and **20** can be merely constructed of elastic ribbon, it has been found preferable to enclose the elastic ribbon within a loose fitting fabric tube **24** where the fabric is smooth to the touch. In this manner, not only is the appearance enhanced but the elastic on being stretched and relaxed cannot pinch the infant's skin.

In use, before the shirt **13** is placed on the child the first part **11** button and loop connectors **19** are grippingly applied to the shirt front. Next, the button and loop connectors **23** are applied to the shirt back. Now the child is dressed and the hook-and-loop connector secured together. The hook-and-loop connection technique not only makes adjustment easy, but is also convenient for diaper changing.

I claim:

1. An infant shirt securing device for use on an infant wearing a diaper, comprising:

- a first part including,
- a first elastic strip folded onto itself forming two arms arranged in a generally V-shape,
- a button and loop connector secured to each end of the strip for releasable interconnection with the front of the shirt, and
- a nonelastic strip having an end secured to the folded over part of the elastic strip and extending oppositely of the arms,
- a second part including
  - a second elastic strip longer than the first elastic strip folded onto itself forming two arms arranged in a generally V-shape, and
  - a button and loop connector secured to each end of the elastic strip for releasable interconnection with the back of shirt; and
- a first half of a hook-and-loop connector secured to a surface of the nonelastic strip;
- a second half of a hook-and-loop connector complementary to the first connector half secured to the second elastic where it is folded onto itself; and
- each elastic strip is enclosed within a separate fabric tube made of a material that has an external surface that is smooth to the touch.

2. An infant shirt securing device as in claim **1**, in which the first elastic strip is approximately 5 inches long and the second strip is approximately 12 inches long.

* * * * *

Trademark Registration

Nº 1511028

THE UNITED STATES OF AMERICA

CERTIFICATE OF REGISTRATION

This is to certify that the records of the Patent and Trademark Office show that an application was filed in said Office for registration of the Mark shown herein, a copy of said Mark and pertinent data from the Application being annexed hereto and made a part hereof,

And there having been due compliance with the requirements of the law and with the regulations prescribed by the Commissioner of Patents and Trademarks,

Upon examination, it appeared that the applicant was entitled to have said Mark registered under the Trademark Act of 1946, and the said Mark has been duly registered this day in the Patent and Trademark Office on the

PRINCIPAL REGISTER

to the registrant named herein.

This registration shall remain in force for Twenty Years unless sooner terminated as provided by law.

In Testimony Whereof I have hereunto set my hand and caused the seal of the Patent and Trademark Office to be affixed this first day of November, 1988.

Commissioner of Patents and Trademarks

PTO–130

Int. Cl.: 25

Prior U.S. Cl.: 39

Reg. No. 1,511,028

United States Patent and Trademark Office Registered Nov. 1, 1988

TRADEMARK
PRINCIPAL REGISTER

RYDER, JUDITH A. (UNITED STATES CITIZEN)
5064 GAVIOTA AVENUE
ENCINO, CA 91436

FOR: BABY'S CLOTHING ACCESSORY IN THE NATURE OF A STRAP TO KEEP SHIRTS IN PLACE, IN CLASS 25 (U.S. CL. 39).

FIRST USE 4-2-1987; IN COMMERCE 4-2-1987.

NO CLAIM IS MADE TO THE EXCLUSIVE RIGHT TO USE "LITTLE SHIRT", APART FROM THE MARK AS SHOWN.

SER. NO. 668,509, FILED 6-25-1987.

MARK TRAPHAGEN, EXAMINING ATTORNEY

Confidentiality Agreement

A confidentiality agreement, as the name implies, states that the signer will not share your idea with anyone else. Before you reveal your idea to a manufacturer, an illustrator, or any other outside party, it's wise to ask them to sign on the dotted line. It is standard procedure when dealing with proprietary ideas and shouldn't be viewed as an indication of distrust; it's simply the proper way to conduct business. You will need to see an attorney to get a confidentiality agreement for your own use, but here is a sample of one I've used in the past.

Dear ______

The undersigned agrees and acknowledges that the documents, drawings, descriptions, and/or any other reproductions, written or verbal, that have been supplied to the undersigned are solely for use in conducting performance and other evaluation tests of a device, and he/she agrees to maintain said drawings and/or other documents or verbal descriptions in confidence and not to discuss this project or constructional or operational details of said project with anyone other than those necessary to conduct evaluation tests, nor will he/she permit anyone to make drawings, copies, or detailed analysis of the device or devices.

The undersigned also agrees and acknowledges that the above referenced project is the sole property of The Company and that by providing the drawings and/or documents to the undersigned for evaluation thereof, The Company does not transfer rights or license to manufacture the device.

Sincerely,

Letter of Intent

A letter of intent is what you might call an agreement to agree. It lets each party know that the other is serious about negotiating a deal, and it gives both the opportunity to put their thoughts in writing so negotiation can begin. A letter of intent can be much more casually written than the one I've provided here, with fewer details, but it can still serve the same general purpose. (This letter of intent became the basis for the actual contract.)

Ms. Judy Ryder
Vice President
Sales and Marketing
American Baby Concepts, Inc.
9574 Topanga Canyon Blvd.
Chatsworth, CA 91311

Re: "New Age Laces"

Dear Ms. Ryder:

We have reviewed the possible application of your lace system to our children's product line and would like to take the next step toward an agreement that would permit us to implement it on a production basis.

We understand from your patent lawyer that you will very shortly be filing an application for U.S. Letters Patent that covers your lace system, and that when this has been done, we will have an opportunity to review it, particularly with regard to the scope of coverage of its claims.

Accordingly, we propose the following arrangement: An exclusive license of the invention from you, renewable annually by us at our sole option, that provides for a royalty of _____ per pair of the laces that we sell in the U.S., its territories and possessions, against a guaranteed minimum annual royalty of _____, payable in advance of each year of the license, including the first.

We would not be obligated to pay you anything more than the _____ advance during any year of the agreement, until such time as your patent actually issues, and then, only the royalty amount accruing on the pairs of the laces that we sell domestically during such year that are *in excess* of _____ pairs ($_____/$_____/pair), which payments would be made on, say, a quarterly basis. This arrangement should avoid our having to pay a substantial royalty on a product that is not protected by a patent, and at the same time, provide a strong incentive for you to get your patent issued in the shortest time possible, in order to increase your income from it.

We do *not* propose to pay you a royalty on any pairage that we sell in countries outside of the U.S., unless you are inclined to seek patent protection in such countries. As you know, this can be an expensive proposition and is a matter that you will want to discuss with your lawyer. We might be willing to assist you with some of these expenses, at least for those countries in which we are fairly sure of being "knocked-off" on this product, but we would expect to be able to use that assistance as an offset against any amounts that we would be obligated to pay you under the license agreement, in accordance with some mutually agreeable arrangement. This is a matter that we can discuss further when we've both had a chance to consider it in more detail.

Thank you for your continued interest in our company.

Sincerely,

License Agreement

A license agreement is a contract between two parties, the licenser and the licensee, allowing one party to use, for a fee, something that is patented or trademarked by another party. The details in a license agreement are usually quite lengthy, covering areas such as how and when payment is to be made, exactly how the company is allowed to use your idea, which party will pay fees associated with patent infringement (someone else using your idea), minimum commitment on the part of the company, and more. Here's part of the draft agreement to license my Little Shirt Anchor to A-Plus Products.

Dear ______:

A-Plus shall make advance periodic payments to Ryder Products of $10,000 per quarter, namely, January 1, April 1, July 1, and October 1 for the following quarter, which payments shall be creditable against actual commissions earned during the quarter.

A-Plus agrees to payment of a commission to Ryder Products based upon net sales of the Product in accordance with the following yearly sales schedule:

(a) $01 – $99,999, 10%;
(b) $100,000 – $249,999, 8%;
(c) $250,000 – $499,000, 7%;
(d) $500,000 – $999,999, 6%; and
(e) One million dollars or more, 4%.

Ryder Products has presently pending two U.S. patent applications on the Product, and upon issuance of any patent thereon Ryder Products and Judy Ryder shall assign to A-Plus all rights, titles, and interest therein for continued payment as set forth herein during the life of the patent upon payment preparation and prosecution costs.

Ownership in the trademark "LITTLE SHIRT ANCHORS" and the United States registration thereof will be transferred to A-Plus and the assignment recorded in the U.S. Patent and Trademark Office.

Judy Ryder agrees to fully identify all vendors of materials and manufacturers used for the Product along with the Ryder Products' most recent pricing structure. It is understood and agreed that Happy Family Products will be contacted first regarding any future orders.

Judy Ryder agrees to act as a consultant for the Product up until August 1, answering all questions relating to the manufacture and marketing of the Product without additional charge to A-Plus. In the event there is need for her service after August 1, Judy Ryder agrees to provide the requested services for a mutually agreed period/s of time and for an agreed upon reimbursement.

During this agreement Judy Ryder or Ryder Products, or their nominee, shall have the right to examine all records of A-Plus that pertain to sales of Products and payments to be made under this agreement.

A significant breach by A-Plus of any terms herein shall give Judy Ryder and Ryder Products the right to terminate upon giving 30 days, written notice to that effect.

Upon termination Judy Ryder and Ryder Products shall have the right to reassignment of any of its patents, the right to purchase any Product inventory at cost, and to receive copies of customer lists and other pertinent documents.

Sincerely,

Product Idea Proposal for a License

Here is a proposal I developed for one of my inventions, which I submitted to companies considering it for licensing. I have replaced some key words with blanks to protect the confidentiality of my product.

Included: Product Description and Purpose
Manufacturing Considerations
Market Testing
Marketing Considerations and Suggested Name
Sizing and Product Line Considerations
Summary
Letters from Users of Product

Product Description and Purpose

The product is a shoelace made of ______ material that resembles a standard shoelace. After being laced in a shoe, this shoelace will attach at the top with a device such as a _____, replacing the need to tie a bow. This enables the shoe to become a "slip-on" shoe, but still allows the preferred appearance of laces. The _____ fastening device may be unfastened to put the shoe on if necessary.

Manufacturing Considerations

The laces would be applied to the shoe. They are laced first with a lace, then a _____ or other fastening device is applied with a machine. Each pair of shoes would require ______ (time) labor, depending on the shoe size and style. Material costs are between ______ (cents). Decorations and adornments would vary in cost.

Market Testing

Market testing was conducted on a paid basis for a three-week period. The test included a wide range of demographics:

_____% of our adult test subjects preferred the test laces to their old laces.
_____% had no preference.
_____% of parents with children under _____ years of age rated the product _____ or higher (on a scale of 1-10, 10 being _____ the highest). These parents cited neatness, conve-

nience, and ____ safety as the best features.

_____% of parents experienced some difficulty in putting the shoes on their child. (This indicates that some styles of shoes such as _____ may not be suitable for these _____ particular laces.)

Children through age _____ cited they looked "cool" or different. Several mentioned how much quicker they are to put on compared to a standard lace. They also mentioned that their friends had noticed and asked about them.

Marketing Considerations and Suggested Name

The name ______ (product name) came through our market test with _____% rating. This name is self-descriptive, short, and lends itself to a number of marketing possibilities. With this product you have the fantastic blend of a utilitarian-type product as well as a ________. Needless to say, kids shoes are a major market. Any increase in market share would be desirable. For adults, just the simple fact that they are more convenient because you don't need to tie the lace since they come with the shoe. The consumer also retains the desired look of a conventional shoelace.

Sizing and Product Line Considerations

Preliminarily, our tests show the shoe designs that are the best candidates for _____ (product name) are designs in which the shoe is ______. These are the same considerations that would be given for any slip-on shoe. Some styles may need to be available in a _____ and _____ sizes. We feel this needs to be researched after some responses to our foucus group testing.

Because these laces require no ____________ to the shoe, if desired, shoe styles could be offered with either a _____ (product name) or a conventional _____. This makes the inclusion of ______ (product name) into a product line very simple and less costly than other designs.

Summary

Initial market testing is very positive. The product serves a definite need while also being a ________ item. It is suitable for ______ age groups. The possibility of exploiting the lucrative kids and young teen markets is great. Implementation of this item into the product line requires nominal expense.

Distribution Agreement

Before you begin working with a distributor you should detail your expectations in an agreement. Here is a sample of one that I have used.

Distribution Agreement

This Agreement is made between XYZ, Inc. of _______________ (hereafter "XYZ"), who's principal place of business is____ and ______________________(hereafter "vendor") who's principal place of business is:____.

WHEREAS, XYZ has and maintains a highly professional sales organization for the purpose of marketing juvenile accessory items to retailers throughout the U.S. and elsewhere; and

WHEREAS, Vendor has a juvenile accessory item, or items, which are called: ____; designed for the purpose of ____, and

WHEREAS, Vendor wishes to engage the services of XYZ to market said product or products in the manner set forth below;

IT IS THEREFORE AGREED AS FOLLOWS:

1. Empowerment: Vendor hereby grants to XYZ the right to market its accessory item(s), (hereafter "Product") on an exclusive basis in the following geographic areas: ____,

(hereafter, "Territory").

2. Trade Segments: The exclusive rights conferred in Para. 1 above shall apply to the following segments of the juvenile product trade:*Juvenile Specialty Stores, Juvenile Catalogs, All Catalogs, Mass-Merchandisers, Hospital Gift Shops, Military PX's.* XYZ is granted authority to market the Product on a non-exclusive basis to the following segments of the juvenile product trade: *Food & Drug Chains and all other segments not specifically excluded in para. No. 4 below.*

3. Territory Exclusions: Geographic territories specifically excluded from this Agreement are as follows:____.

4. Trade Segment Exclusions: Market segments specifically excluded from this Agreement are as follows:____.

5. Mutual Obligations: Vendor agrees to sell and XYZ agrees to buy sufficient product at prices agreed upon between the parties to allow XYZ to make prompt shipment to its accounts. Nothing herein shall be construed to compel XYZ to buy any specific amount of product from vendor.

7. XYZ Obligations: XYZ agrees to use its best efforts to promote the sale of the product throughout the Territory to all retail segments for which marketing rights have been granted under Para. 2 above (subject to exclusions in Para. 4 above); to maintain professional sales representation throughout the Territory to the best of its ability; to provide samples of Product to each such sales representative; properly inform them of the products' features and benefits; and manage their efforts and maximize sales of Product.

8. XYZ's Functions: XYZ's responsibilities shall include credit approval of all retail accounts, approval of orders, shipment on orders, extension of payment terms to accounts, collection of accounts, and all other responsibilities commonly assumed by other distributors in trade practice.

9. Vendor Warranties: **VENDOR WARRANTS TO XYZ THAT ALL PRODUCTS(S) SHIPPED TO XYZ PURSUANT TO THIS AGREEMENT ARE FREE OF DEFECTS IN MATERIAL AND WORKMANSHIP; THAT THEY COMPLY FULLY WITH ANY AND ALL SUGGESTED OR REQUIRED SAFETY RECOMMENDATIONS, REQUIREMENTS OR LAWS, AS PROMULGATED BY ANY LAW MAKING BODY, CONSUMER PRODUCT SAFETY COMMISSION, THE FOOD AND DRUG ADMINISTRATION, AND ANY AND ALL OTHER GOVERNMENTAL OR NON-GOVERNMENTAL AGENCIES OR GROUPS; THAT THE PRODUCT IS FIT FOR THE PARTICULAR PURPOSE FOR WHICH IT IS MARKETED AND FOR THE PERSONS IT IS TO BE MARKETED TO; THAT PRODUCT IS STATE OF THE ART AND THAT IT MEETS INDUSTRY STANDARDS, IF ANY, FOR QUALITY, PERFORMANCE, SAFETY AND DURABILITY; AND ALL OTHER WARRANTIES UNDER THE UNIFORM COMMERCIAL CODE.**

10. Indemnification: Vendor agrees to indemnify and save XYZ harmless for any and all loss, cost or damage on account of any injury to persons or property occurring in the performance of this Agreement, except as caused by XYZ's negligence.

11. Insurance: Both parties agree to insure their own interests and neither shall rely upon the other for this purpose.

12. Effective Date: The effective date of this Agreement shall be the date upon which XYZ signs, as indicated opposite its signature below, and shall continue for a period of two (2) years after such date, automatically renewing for like terms on each anniversary date, or until terminated as provided below.

13. Termination by Vendor: This Agreement may be terminated by vendor by declaring XYZ in default in the event XYZ commits a material breach of this Agreement, provided written notice is given XYZ of the alleged default giving XYZ 60 days to cure said default. In the event the default is cured, the notice of termination shall be rendered null and void and the Agreement shall continue in force as if a default had not occurred. In absence of default or cure by XYZ, Vendor may terminate this Agreement at any anniversary date by giving XYZ notice in writing not less than 90 days prior to said anniversary date, to be effective at the end of the term.

14. Termination by XYZ: This Agreement may be terminated by XYZ by giving 60 days notice in writing to Vendor, however, said notice and subsequent termination shall not relieve XYZ or vendor of any responsibilities incurred to the effective date of termination, including, but not limited to, payment of all moneys due under the Agreement, or the continuation of shipment by vendor to XYZ during the notice period.

15. Repurchase: In the event of termination or expiration of this Agreement, unless immediately renewed by mutual agreement, Vendor agrees to immediately repurchase all product then in XYZ's inventory which is in salable condition at the price XYZ paid vendor for said product, F.O.B. Wheatland, Iowa and to accept said product on a "collect basis".

17. Assignment: This Agreement may not be assigned without the express, written approval of the other party and vendor agrees not to sell or otherwise transfer ownership or control of the Product to a third party during the term of this Agreement unless such transfer of ownership or control is made subject to the continuation of this Agreement.

18. Pricing: The prices to be charged XYZ for the product may not be increased without first giving XYZ 60 days notice and allowing XYZ the option to take delivery of any quantity of Product at the current price before increase. In no event may Vendor increase the price charged XYZ more than 5% during any one twelve (12) month period of this Agreement.

19. Changes to Agreement: No changes shall be made in the terms of this contract except in writing, signed by both parties, and appended hereto.

20. Validity of terms: In the event any term, condition or provision contained in this Agreement shall be declared null and void through a judicial proceeding or process, in a court of competent jurisdiction, this Agreement shall be divisible of such term or condition and the remaining terms and conditions shall remain in full force and effect.

21. Venue: This Agreement shall be deemed to be made in Wheatland, Clinton County, Iowa and any judicial action instituted by either party to enforce the terms of this Agreement shall be commenced in the District Court of the State of Iowa in Clinton County. All costs, including reasonable attorney's fees, incurred through any court proceeding or other action arising out of this Agreement shall be paid entirely by the losing party.

AGREED on the date last written below.

Date:________________ ______________________________

Vendor Company Name

By:______________________________

Date:________________ XYZ, Inc.

By:______________________________

Sales Rep Agreement

When you hire sales reps to sell your product to buyers, you will want to spell out the specifics of your relationship. Here's a sample agreement.

This agreement is entered into this _______ day of _________ 19___ by and between The Company and ________________, whose mailing address is _______________.

Products

The products for which Salesperson shall act as Sales Representative are described on Schedule 1, attached to this Agreement.

Territory

The Territory means the area described on Schedule 2 attached to this Agreement. It shall include not only geographic territory but also the trade segments Salesperson is authorized to solicit.

Appointment as sales representative

Company hereby appoints Salesperson as its Sales Representative to solicit orders for the Products in the Territory. Salesperson hereby accepts the appointment and agrees to solicit orders and promote the Products, subject to the terms of this Agreement.

Orders

Orders for the Products received by Salesperson shall be immediately forwarded to Company. Salesperson shall use forms and documents provided to it by Company, or approved by Company, for such orders. Each order solicited by Salesperson is subject to acceptance by Company, and Company may reject any order, in whole or in part, for any reason, including without limitation, lack of availability of inventory or supplies or limited capacity of Company, or creditworthiness of the customer submitting the order by Salesperson accompanied by the required information.

Warranties

Salesperson shall not make any representations as to warranties to any customer or potential customer other than the warranty published by Company, if any, in its sales literature or in its Sales Order from time to time.

Salesperson agrees to:

a) use his/her best efforts to sell and promote the sale of the Products within the Territory and to develop a volume of business of Products commensurate with the potential for sales of the Products in the Territory and to abide by Company's policies as communicated from time to time

b) contact and solicit prospective purchasers of the Products in the Territory, make demonstrations to promote sales, and promptly respond to all inquiries for the purchase of the Products

c) devote such time as may be necessary for the purposes of soliciting or promoting the sale of the Products

d) represent no other products that compete with the Products without prior written approval of Company

e) maintain a sales office in the Territory

f) assist Company in the collection of amounts owed by customers for Products sold on credit to the extent reasonable

g) submit to Company such reports as it may request from time to time.

Company agrees to:

a) establish the price, charges, and terms and conditions of sale of the Products

b) furnish Salesperson with current price lists, terms, quantity discounts and freight charges and credit applications for customers

c) provide its customary literature and samples concerning the Products to Salesperson or to customers as Salesperson may from time to time request.

Commissions
a) Company shall pay to Salesperson a commission on the sale of the Products solicited by Salesperson in the Territory ____% of the "net sales price." The "net sales price" shall mean the wholesale price of the product sold less all returns, discounts, allowances, and freight charges.

b) Salesperson's commission shall be earned when the invoice for the Products sold is sent to customer. The commission shall be paid no later than the 25th of the month following the month of the invoice.

Relationship of parties
Salesperson's relationship to Company in the performance of this Agreement is that of an independent contractor. The personnel performing services shall at all times be under Salesperson's exclusive control and direction and shall be employees of Salesperson and not employees of Company. Salesperson shall cover or insure all of his/her employees, if any, performing services in compliance with applicable worker's compensation or employer's liability insurance laws. Salesperson shall not have authority to bind Company in any manner.

Miscellaneous
a) Salesperson shall provide Company with a complete list of lines, products, or goods represented by Salesperson and shall notify Company within ten (10) days of the date Salesperson adds any new lines, products or goods.

b) Salesperson shall not assign its interest under this Agreement without prior written consent of Company.

Termination
This Agreement shall become effective upon its execution by the parties. The Agreement may be terminated by either party, with or without cause, by giving the other party thirty (30) days written notice. Company will pay commissions on all orders submitted by Salesperson to, and accepted by, Company prior to the date of termination in accordance with ____________.

Governing law
The parties agree that this Agreement shall be governed by the laws of the State of ________, as applicable to contracts to be performed entirely within that state.

Press Release

FOR IMMEDIATE RELEASE

October 1, 1994

Contact: Mary Smith
1-800-234-FLAP

Flap Happy Ranked 194 on Inc. magazine's 13th Annual Listing of the 500 Fastest-Growing Private Companies in America

Santa Monica, California—In a special annual issue, *Inc.* magazine has announced the *Inc.* 500, a ranking of America's 500 fastest-growing private companies. Flap Happy in Santa Monica is ranked 194 on the 13th annual list. The special *Inc.* 500 issue goes on sale Tuesday, October 18, 1994.

Flap Happy was founded back in 1987, when Laurie Snyder decided to design a hat to shade her child's face from the sun, thus, designing the first Flap Hat, with enlarged brim and "camel flaps." Today, Flap Happy produces hats in several styles and fabrics, as well as an extensive clothing line in sizes newborn to 10. Some hats even come in adult sizes. Flap Happy products are now available throughout the United States as well as Canada and parts of Europe. All items are made in the United States from quality materials and exceptional workmanship.

For further information, please call 1-800-234-FLAP or stop by the Flap Happy Factory and Outlet Store located at 1714 Sixteenth Street, Santa Monica, CA 90404.

To be selected as a member of the 1994 *Inc.* 500, a company must be independent and privately held through the 1993 fiscal year; must have shown at least $100,000 in 1989 sales, but no more than $25 million; and must have shown a sales increase between 1992 and 1993. Rankings are determined as a function of percentage sales growth over the period 1989–1993.

Members of the working media with further questions concerning the 1994 *Inc.* 500 may call Tom Butler at 212-843-8065.

Press Release

FOR IMMEDIATE RELEASE

September 1, 1994

Contact: Mary Smith
1-800-234-FLAP

Flap Happy is proud to announce an innovative line of clothing made from a remarkable new fabric derived from *recycled plastic bottles*. Sound ridiculous? Not to environmentalists.

Researchers recently discovered the perfect solution to the serious problem of landfill overflow by creating a comfortable, attractive fleece made from recycled soda bottles. Not only is ECOFLEECE soft and warm, it will relieve the nation's landfills of some of the 60 million bottles discarded each day.

ECOFLEECE is indistinguishable from other high-quality fleece fabrics. Even after numerous washings, the fleece maintains high standards for strength, shrinkage, and colorfastness.

Sportswear made from ECOFLEECE can be found in the Flap Happy 1994 Fall/Winter Catalog. ECOFLEECE is available in rich shades of plum and loden green.

ECOFLEECE. Clothing that protects your child and your planet.

Flap Happy products are available at stores nationwide. Ask your local baby store or telephone 1-800-234-FLAP.

Collection Letters

When an invoice is 45 days overdue, you should send a letter or call that account. Don't back off. Persistence pays. Here are some examples of collection letters that you may use or borrow from in creating your own.

First Letter:

Dear ________:

Just a reminder, but . . .

HAVE YOU FORGOTTEN SOMETHING?

No doubt at one time or another, we've all had occasion to be thankful for being reminded of something we've overlooked.

This letter is a friendly "tap on the shoulder" to remind you of your unpaid account of $________. Your check, as yet, has not arrived. We know you'll appreciate being reminded.

Just put your check in the return addressed envelope and please mail it today. If this is not possible at this time please call me so that other payment arrangements can be made.

Sincerely,

Final Letter:

Dear ________:

I'm happy to say that it's seldom necessary for us to turn an account over for collection. When circumstances leave us no choice, the customer is always informed of our intended action.

Neither you nor I want to see this happen. We pride ourselves on the friendly relationships we have established with our customers and a lawsuit would destroy that. We value your account, and are exceedingly reluctant to take any step that would eliminate that goodwill.

You will surely agree that we have been fair and most patient. However, the time has come when YOUR ACCOUNT MUST BE PAID. Unless you show some effort to protect your credit rating, I have no choice but to authorize collection action.

Please mail us your check immediately so that you can avoid this final step.

Sincerely,

Request for Product Information

REQUEST FOR PRODUCT INFORMATION

NOTE: It is imperative that a copy of this completed fonn be signed and returned to The Right Start, Inc. in order for an item to be considered.

Right Start Plaza
5334 Sterling Center Drive,
Westlake Village, CA 91361
TEL 818•707•7100
FAX 818•707•7132

MANUFACTURER'S INFORMATION

Company Name: ____________________

Contact: ____________________

Street Address: ____________________

City/State/Zip: ____________________

TEL ____________ FAX ____________

SALES REPRESENTATIVE INFORMATION

Rep Firm Name: ____________________

Contact: ____________________

Street Address: ____________________

City/State/Zip: ____________________

TEL ____________ FAX ____________

R.S.I. office use only:

Date Rec'd ____________

Sample Rec'd ____________

SKU type (C/R/B) Other: ____________

Dept ______ Sub ______

P/L ______ TX ______

Excl ____________

O/S ❑ Drop Ship ❑

Ins VBF ❑ Exp: ____________

ID# (1) ____________

ID# (2) ____________

Retail $ ____________

LBS ____________

We are considering the item named below for use in our catalog and/or retail stores. However, we must have completed information before making a final decision. Please complete this form including date and signature and return to us no later than ____________

NAME OF ITEM ____________ YOUR STOCK NUMBER ____________

PRICE AND DELIVERY INFORMATION

	DOMESTIC	IMPORT
1. Suggested Retail Price		
2. Wholesale Cost (before Catalog Allowance)		
3. Catalog or Advertising Allowance		
4. Other Discounts (Qty)		
5. Net Wholesale Cost		
6. Payment Terms		
7. Freight Allowance		
8. Actual Range of retail price nationally		

9. When will this item be available to ship? ____________
10. Is this a continuing item in your line? ____________
11. Will your item be available until ______? ❑ Yes ❑ No
    a. Will you maintain your price for this period? ❑ Yes ❑ No
12. Shipping time you require from receipt of our order
    a. Domestic shipments: ______ days
    b. Import shipments: ______ days
13. Product shipped from
    a. Domestic ____________
    b. Import ____________
14. Minimum quantity requirement ____________

PACKAGING INFORMATION

1. How is the individual item packaged? (check one)
   bag ❑ box ❑ loose ❑ other ❑
2. Package Dimensions: H: ____ W: ____ D: ____
3. Package Shipping Weight ____________
4. How pieces to a Master Carton ______ Lbs ______
5. Case Pack Count ____________
6. Retail packaging available? ❑ Yes ❑ No
   a. Different price? ____________
7. Re-shipper pack available? ❑ Yes ❑ No
8. Retail display available? ❑ Yes ❑ No
9. Private label available? ❑ Yes ❑ No
   comments: ____________
10. Does package meet Postal and UPS maximum weight and dimensions? ❑ Yes ❑ No
11. UPC bar codes available? ❑ Yes ❑ No

PRODUCT INFORMATION

1. Product Dimensions: H: ____ W: ____ D: ____
2. Materials: ____________
3. Finish / Colors: ____________
4. How long has this item been on the market? ____________
5. If electrical, is it U.L. listed? ❑ Yes ❑ No
6. Does this item meet all applicable U.S. safety standards and regulations? ❑ Yes ❑ No
7. Please list applicable certifications and/or testing information: ____________
8. Please list any awards and/or endorsements: ____________
9. Age range: ____________
10. Weight range: ______ Maximum weight: ______
11. Will this item be sold exclusively through The Right Start? ❑ Yes ❑ No
    a. If so, for how long? ____________
12. List key consumer benefits this product offers: ____________

SPECIAL INFORMATION

1. Do you produce this item yourself? ______ Is it imported? ______ Country of Origin? ______
2. Please send us the following: ❑ Color transparency ❑ Sample/Prototype ❑ Product literature
   ❑ Vendor's Broad Form Liability Endorsement ❑ Suggested ad copy ❑ Line art for award/certification logos
   ❑ Claim substantiation (ie. testing and/or certification literature)

1. All information contained herein will be used for the purpose of offering your merchandise in our retail stores and/or catalogs. Therefore, we will hold you responsible for all information furnished and for price and delivery times quoted. Information provided may be used in the advertisement of your product. Your signature below indicates that all facts provided are accurate.

2. By your signature below, you agree that no price increases shall be effective without 60 day written notice to The Right Start of the new price and effective date. New models or style changes and substitutions must be shipped at the old model price unless 60 days prior notice is received.

3. All orders must be shipped within 20 days of Purchase Order date or be shipped prior to the Purchase Order cancellation date, whichever is later. A late shipment penalty of 5% for each 5 business days will be deducted off invoice on all late shipments.

Signature ____________ Company ____________

Title ____________ Date ____________

GLOSSARY

Acquisition When one company takes over ownership of another.

Advertising Promoting a product or service, by paying for time on television or space in printed materials such as magazines.

Back order The items that are not included when an order is only partially filled and shipped, usually because they are out of stock temporarily.

Blister packaging A printed card that is fairly rigid, made with a special coating, onto which a plastic bubble, or blister, is attached that contains a product.

Blow molding A manufacturing process for plastic items in which air is used to distribute plastic into the contours of a mold, thus creating a hollow plastic product.

Bulk Merchandise that is not in a display package.

Business plan A written document that encompasses all phases of a new business, demonstrating the amount of expected growth, how it will be achieved, use of capital, cash flow projections, market synopsis, etc.

Capital Money used to start, expand, or operate a business.

Catalog sheet A one-page printed display of a product or group of products used to promote the item(s).

C.O.D. (Cash On Delivery) A delivery of merchandise that must be paid for at the time delivery is made.

Commission A percent of the actual dollar amount of a sale that is paid to the salesperson as compensation.

Components The parts needed to make a product.

Component sourcing Finding suppliers for the parts needed to make a product.

Confidentiality agreement A written agreement that, when signed, binds the receiving party to keep the disclosed information confidential.

Consumer An individual that will potentially purchase a product for his or her own use.

Co-op advertising An advertising program that is paid for jointly by the manufacturer and the retailer in order to promote the manufacturer's products sold in the retail outlet.

Copyright Protection offered by law for literary property, such as a song or book, to deny others from using it for gain.

Customer base All the customers a company has ever sold to.

Customer service The service provided by a company to answer questions, handle complaints and otherwise deal with its customers.

D.B.A. (Doing Business As) is a fictitious name statement for a solely owned business or partnership, that must appear publicly before commencing in business. (For example, Harry's Flower Shop is a D.B.A. for the owner, Harry Johnson.)

Direct mail Solicitation of sales done directly through the mail to reach targeted customers.

Distribution The system by which a manufacturer gets its product to the consumer.

Distributor A company or individual that markets or delivers products from the manufacturer to the retailer or consumer.

Editorial The portion of a publication that is not a paid advertisement, such as articles, features, and columns that express opinions and other non-news items.

Endorsement A statement in support of a product or service, usually from a noteworthy individual or organization or expert in the field, that lends credibility to the product or service and induces persons or businesses to buy.

Exclusive (Rights, Agreement, etc. . . .) All rights to a patented, copyrighted, or trademarked item.

Expenses Costs associated with conducting business.

Export To send products from the U.S.A. to a foreign country.

Extrusion molding A manufacturing process for plastic in which plastic is pressed or forced through a mold to retain a specific shape throughout its length.

Financing Money that is borrowed to start, expand, or operate a business.

Focus group A preselected group of people who are placed in a controlled environment, then asked to give their opinions about a product or service.

Footprint The amount of floor space taken up in a store by an object, such as a product display.

Freight, costs (expense) The shipping charges to send a product from one place to another.

Freight-in The shipping charges a product owner incurs, to get his or her products or materials from suppliers. Also referred to as back freight.

Freight-out The shipping charges a customer incurs to get products from the manufacturer or distributor.

Gross In financial terms it means the total of something, such as dollars in sales, prior to deducting any expenses.

Gross profit margin The amount over the actual cost of an item that a company sells the item for.

Hang tag A tag or card that is attached to an item by a thin length of plastic.

Header card A printed card that is usually folded in half and attached to a product or polybag containing a product.

High credit The highest amount of credit a company has ever extended or is willing to extend to a particular company or individual.

Illustration A drawing, as opposed to a photograph.

Import To bring products or components into the home country (i.e., U.S.) from a foreign country.

Infringement Refers to the rights of an individual or company being violated by the actions of others—usually associated with patents, copyrights, or trademarks.

In-house Something that is done within a company by company personnel, without the assistance of outside companies or individuals.

Injection molding A manufacturing process for plastic in which plastic is injected into a mold to form a specific, solid shape.

Inventory The products and components a company has on hand for sale to others.

Invoice An itemized bill for product(s) or services.

Invoicing Generating an invoice.

Keystone A term used in wholesale that refers to doubling a price.

Knock off When one company copies another company's product or idea for their own gain.

Labor The actual handling required to make, package, or ship a product.

Letter of intent A letter from one party to another, which may have specific legal consequences, expressing both parties' intention to proceed in a specific matter.

License A document used to detail the terms under which an individual or company allows another individual or company certain rights to an exclusive item covered by patent, trademark, or copyright.

Licensee The individual or company that receives rights to an item covered by a patent, trademark, or copyright granted by a license.

Licenser The individual or company that grants rights to an item by license.

Margin The percentage a company sells a product that is over the cost of that product. Gross margin is before expenses; net margin is calculated after expenses.

Market test A process used to evaluate the market's response to a product or sales strategy before it is offered for sale or use, to prove its appeal or effectiveness.

Mass merchant Multiple retail establishments, under one ownership, that cater to several different markets, such as jewelry, women's clothing, housewares, gardening, pets, etc. to the public, such as television or newspapers.

Media Modes of communication that transmit information directly

Merchandising Sales and promotion plans, strategies, and efforts, usually at the retail level.

Merger When two or more companies combine to make one.

Net sales This term varies in meaning depending on use, though it implies total sales minus specific deductions such as returns, defective items, cost of sales, etc. When used it must be clarified. It can be used for a dollar amount or an item total.

Net terms This indicates the amount of time after the invoice date in which an invoice must be paid—for example, net 30 means it must be paid within 30 days of invoice date.

Non-confidentiality agreement An agreement that specifically states that at least one party signing cannot be bound to keep the disclosure in confidence, and that all parties are aware of this.

Overhead Set costs for operating a business that do not vary greatly from month to month, such as salaries, utilities, advertising, etc.

Patent Protection offered by law for a product idea, to deny others from using it for their own gain.

P.O. See Purchase Order.

Point of purchase (P.O.P.) A sales tool such as a display, set up in a store where the featured product can be purchased.

Premium An incentive used in sales, usually as an item offered free with the purchase of another item to induce buying by increasing the perceived value.

Prepaid freight A term indicating that the company selling the product will pay the freight to ship the product for the company purchasing it.

Press release Written material, often accompanied by photos or illustrations, supplied to the media to inform them about an event, product, person, etc.

Product liability The legal responsibility a product manufacturer, distributor, supplier, or seller has regarding goods they handle.

Product liability insurance Insurance policy to cover product liability claims.

Product line All products offered by one company, or a specific category of product offered by one company.

Promotion Using methods other than paid advertisements to promote the sale and popularity of a product—for example, brochures, coupons, free samples, public relations, etc.

Prototype An actual sample of a new product, used to determine effectiveness, manufacturing plans, size considerations, etc.

Public relations Communication with the media regarding a company, a person, products, or services, with the aim of interesting the media in broadcasting or publishing the information supplied, thereby promoting the person, product, or services to its audience.

Purchase order An itemized order sent to a company to purchase products.

Quality control Maintaining manufacturing standards so that the quality of the product being manufactured will always be the same.

Quantity discount A discounted purchase price of an item when buying a large number. Also referred to as a volume discount.

Quote Details of a sales transaction, such as price, delivery, etc., offered prior to a transaction actually taking place.

Retail The sale of goods sold to the end user, or a description used for a business that sells to the public.

Revenue Monies that a company collects while conducting business.

Sales projections A detailed estimate of future sales.

Sales representative An individual or organization that represents a company by soliciting sales for the company.

SBA See Small Business Administration.

Single-product company or vendor A company that offers only one product for sale.

Small business administration (SBA) An agency of the U.S. government designed to be a resource and offer assistance to small businesses. There are regional offices in every state.

Source To find a supplier for a product or material.

Specialty store A retail establishment that caters to one specific market.

Statement A summary of all invoices due and payments received, which is generated by the seller and sent to the buyer, usually monthly.

Supplier A company that sells materials/products to other companies.

Terms The conditions under which a sales transaction is achieved. Usually refers to payment, delivery, price, cancellations, etc.

Territory A geographic or otherwise defined area, usually referred to when discussing sales responsibility.

Time study Used by a manufacturer when determining the length of time it takes to make a certain product or a particular procedure when calculating price.

Trademark Protection offered by law for a word or phrase, usually used in conjunction with a specific symbol, such as a name, logo, or slogan.

Typesetting The process of putting words in the exact place and in the exact size and type style in preparation for being printed.

Vendor A company that sells products to another company.

Venture capital Money that is invested in a company, either to start or to expand, by someone willing to speculate on the company's future success.

Volume discount See Quantity Discount.

Wholesale The sale of goods from the manufacturer or the distributor to a retailer. Ca also refer to price in the same transaction.

RECOMMENDED READING

Many of the titles marked with an ★ may be hard to find in libraries and bookstores; they are, however, available through The Investor's Bookshop catalog (see below).

★*How to Design Better Products for Less Money*, A.V. Gunn, Publisher: Halls of Ivy Press.

★*Patents, Getting One, A Cost-Cutting Primer for Inventors*, Stuart R. Peterson, Publisher: Academy Books.

★*Patent It Yourself*, David Pressman, Publisher: Nolo Press.

★*How to Copyright Software*, J. J. Salone, Esq., Publisher: Nolo Press.

★*How to Sell Your Own Invention*, William E. Reefman, Publisher: Halls of Ivy Press.

★*Inventors Journal*, Melvin L. Fuller, Publisher: M&M Associates.

★*Cash In Your Bright Ideas*, George Siposs, Publisher: Siposs Publishing.

★*Marketing, Researching and Reaching Your Target Market*, Linda Pinson and Jerry Jinnett, Publisher: Out of Your Mind and Into the Marketplace.

★*How to Prepare Patent Applications*, John R. Flanagan, Publisher: Patent Educational Publications.

★*Inventor's Marketing Handbook—A Complete Guide to Selling and Promoting Your Invention*, Reece A. Franklin, Publisher: A.U.A. Publishing Co.

Marketing Your Invention, Thomas E. Mosley Jr., Publisher: Upstart.

Patents, Trademarks, Copyrights and Trade Secrets: A Legal Guide to Protecting Your Intellectual Property, John R. Flanagan, (videotape and audiotape).

★*Licensing—A Strategy for Profits*, Edward P. White, Publisher: Licensing Executives Society.

★*Complete Guide to Making Money with Your Inventions*, Richard E. Paige, Publisher: M&M Associates.

Home Made Money, Barbara Braabec, Publisher: Betterway Publications.

★*The Home-Based Entrepreneur*, Linda Pinson and Jerry Jinnett, Publisher: Out of Your Mind and Into the Marketplace.

★*Anatomy of a Business Plan*, Linda Pinson and Jerry Jinnett, Publisher: Out of Your Mind and Into the Marketplace.

★*Recordkeeping: The Secret to Growth and Profit*, Linda Pinson and Jerry Jinnett, Publisher: Out of Your Mind and Into the Marketplace.

★*The New Small Business Survival Guide*, Bob Coleman, Publisher: Norton.

★*Secrets of a Successful Entrepreneur: How to Start and Succeed at Running Your Own Business*, Gene Daily, Publisher: K&A Publications.

From Workshop to Toy Store, Richard Levy and Ronald Weingartner, Publisher: Simon & Schuster.

★*Bulletproof News Releases—Help at Last for the Publicity Deficient*, Kay Borden, Publisher: Franklin Sarrett Publishers.

★*Marketing Without Advertising—Creative Strategies for Small Business Success*, Michael Phillips and Sally Rasberry, Publisher: Nolo Press.

Invention & Evolution—Design in Nature & Engineering, M. J. French, Publisher: Cambridge University Press.

How to Write a Business Plan, Mike McKeefer, Publisher: Nolo Press.

Getting to Yes—Negotiating Agreements Without Giving In, Fischer and Wry, Publisher: Penguin Books.

Other good resources:

The Dream Merchant Newsletter, "The Helpful Hand for Inventors & Entrepreneurs," Torrance, CA

The Lightbulb, "The Magazine of Creativity, Invention, and Entrepreneurship"

Inventors Bookshop
Phone: 916-468-2282
Fax: 916-468-2238
P.O. Box 1020
12424 Main Street
Fort Jones, CA 96032

An invaluable source for books and tapes for inventors. If they don't have it, they'll find it! Their catalog includes all those titles above marked with an asterisk.

INDEX